EDINBURGH
ISLAND BIOLOGY SERIES

Living with Seabirds

BRYAN NELSON

EDINBURGH UNIVERSITY PRESS

Edinburgh University Press
22 George Square, Edinburgh

Set in Linotronic Plantin
by Speedspools, Edinburgh, and
printed in Great Britain by
Butler and Tanner Ltd,
Frome, Somerset

British Library Cataloguing
    in Publication Data
Nelson, Bryan
Living with seabirds—
    (Island biology series)
1. Sea birds
I. Title   II. Series
598.29′24′0924   QL673
ISBN 0 85224 523 8

Contents

Preface

Like Ronald Lockey, William Beebe and, in a different way, Gavin Maxwell, Frank Darling acted as a direct channel through which nature flowed freely. His simple style left plenty of scope for his reader's imagination and, like an audience that participates instead of merely listening, they responded eagerly. The golden thread was his observations, his science and his adventure.

In this book I am trying to do a little of the same thing. In it there are wild places and mild adventure – nothing heroic – but mainly there is the simple pursuit of seabirds. Like Frank Darling and Bobby, June and I lived on small islands, desert or otherwise, studying seabirds. There were a few ideas, a lot of evidence and the painful struggle to make sense of it. In among, we lived in a way that many people dream of doing, waking up every day to another 'free day', no boss, no timeclock, no telephone, no commuting, none of the usual pressures. Just sun, fresh air, interesting wildlife and peace. But it wasn't quite like that. In fact it wasn't at all like that, as I believe the pages of this book will show.

Why was this book written? I was told what not to write. It is not popularised biology. There is a lot of bird behaviour and ecology in it but it is not an attempt to gloss up these subjects. Nor, I devoutly hope, will readers of any of my previous books find much evidence of re-hash. It is often anecdotal, derived from my general diary rather than my field records. Where this sparks off discussion of behaviour, ecology, evolution or anything else, that seems to me appropriate. It did so in my own mind and I hope it will do so for the reader. Because it is intended for the general reader who likes wildlife and wild places and who enjoys the human side of this sort of life, as well as for the more dedicated naturalist,

I have been free to treat the personal aspects in some detail, as indeed Darling did.

Despite the hard slog, living and working on remote sea-bird islands is a most privileged and rewarding way of life. We would not exchange our years on the Bass Rock, the Galapagos Islands and Christmas Island. I hope to convince the reader that, like Darling's, they were productive as well as happy.

ACKNOWLEDGEMENTS

There is really only one person who was indispensable to the activities detailed in these pages. It is my wife. I think she knows that I honour her for it. But life throws up numerous benefactors and I have had more than my share. Some of their names appear in these pages but their lineage extends back to the days of Hector Boece. And writing of ancient lineage, it is a pleasure to thank Sir Hew Dalrymple, first for permission to work on the Bass and then for his continued interest in the gannets. Edna Bremner, of the Zoology Department, Aberdeen, kindly typed much of the manuscript, and I tender my thanks to the University for its support of my work.

1

Beginnings

Hearken, thou craggy ocean pyramid!
Give answer from thy voice, the seafowls screams,
When were thy shoulders mantled in huge streams?
When from the sun was thy broad forehead hid?
John Keats, *Ailsa Craig*.

We haven't yet fallen off the Bass cliffs though on occasion
we deserved to. I shudder now at my careless disregard of
elementary precautions, born of a mixture of laziness and
perhaps a psychological need to create minor crises for the
reward of overcoming them, a dangerous combination. But
fieldwork in lonely and difficult places can be hazardous.
June came within a whisker of a fearful accident on Christmas Island when one of the rungs, which I had nailed to a
tree to help us climb to a nest of Abbott's booby high in the
canopy, split and gave way. The tree stood amongst jagged
limestone pinnacles but by great good fortune the bare bole
of a slim sapling, racing towards the sky, was growing alongside. She fell against this and slid helplessly down its length,
landing softly in the only patch of bare earth amongst a lethal
array of sharp rocks. Had she fallen backwards our venture
and probably her life would have ended then. There it was
again – I had not made sure those rungs were totally safe.
Even after that, I regularly risked life and limb – high in dead
trees after the nests of Andrew's frigatebirds. I knew that it
was risky but I still did it. Nowadays I wouldn't.

I am not a rock-climber but I can see that climbing discipline, like that of the sea, effectively overcomes that fatal flaw,
the tendency to take a risk – a subtle adversary, biding its
time and striking when one is weakest. I still fear it because
no amount of preparation can cover all the minor crises that

I

can arise and it is the split-second decision that counts.

I remember, as a 21-year-old, trying to climb part of the lower cliffs on Ailsa Craig from their boulder base to ring young fulmars and kittiwakes. The fulmars spat their foul-smelling oil into my face whilst the kittiwakes split the air with their impassioned 'kitti-waak, kitti-waak'. At one point, I stuck, and the thought flashed into my mind that perhaps these smells and sounds would be my death knell. At such fragile moments one may give way to a desperate reflex escape act rather than waiting to calm down. The same feeling has recurred in seabird colonies, at fraught moments. One evening in late March I was down the east cliffs on the Bass. I had descended, alone and unroped, to look for early eggs. Nobody knew I was there. It was a grey dusk with a bitterly cold wind and beginning to feel eery. I wanted to be back at the lighthouse fireside and started up the only easy route to the top but must have entered the wrong crack, for it petered out. Suddenly it seemed urgent that I get back up before darkness really fell although in fact there was still enough light. Then came the urge to take a chance rather than go back down and start all over again. Nowadays I wouldn't climb around like that, but I still recognise the weakness, the urge to cut corners. These experiences, though, are part of the aura of seabird colonies and it was Ailsa Craig that really hooked me.

These days, research topics have to be closely defined before they have any chance of attracting a grant from the relevant Research Council. But in 1960 I knew merely that I wanted to study gannets. Seven years earlier a friend and I spent a week on Ailsa Craig. As usual, it was a shoestring affair – about the most expensive item were the rings to put on the gannets' legs – and we had to buy them out of our own pocket. We arrived in Girvan on a hot Saturday evening in July and the following day crossed to Ailsa in the trawler, *Selina II*. It was a rough day and I don't think she really had any intention of landing us. When we reached Ailsa after a choppy passage she turned round and came back. What an anticlimax it was, unloading all our gear again at the harbour. But the next day all was well and we were put ashore at the old jetty on Ailsa. Jimmy Girvan, like his father before him, quarried the beautiful grey-green red-veined granite for curling stones and his wife ran a teashop in a cool, shady barn at

2

Ailsa, 'an iyl, ane myle lang, quherein is ane grate high hill, round and roughe, and ane heavin [haven], and als abound-ance of Soland geise.' (Donald Munro, 1549, *Description of the Western Isles of Scotland*)

the head of the jetty. They kept a herd of goats which browsed on the short, sweet sward and on seaweed, and provided delicious milk for the teashop. Their cosy little cottage, later deserted, became home for Sarah, my PhD student, for three arduous seasons. That same evening we climbed to the top of Ailsa, on the steeply exposed slopes above the awesome cliffs, in the sun and wind, amidst the wheeling, crying gulls and above all, the magnificent gan-nets. Ailsa forged the link.

Kennedy's Nags, chocolate, guano and sheep ticks mean Ailsa Craig and gannets to me. On a blazing day we climbed from sea-level to more than a thousand feet through tick-in-fested bracken, before descending gingerly to the sun-baked ledges amongst the rotten slabs called Kennedy's Nags. There, in the heat and overpowering stench, we tenderly placed our expensive rings on the chicks' dirty black legs, sustaining a criss-cross of deep lacerations on the backs of our hands, for although their beaks lack the steely hardness

3

of the adults' the tips are sharp. Both then and time and again later we found that despite a coating of gannet muck and sulphurous ooze our cuts healed quickly and cleanly, perhaps disinfected by the ammonia and sulphur. Around our ankles and mid-riffs were great red belts of bites from the ticks which swarmed in the bracken. The clegs were jabbing us with red hot needles and itchy gannet fluff stuck to our sweaty faces. But we thought it was great.

The chocolate was for lunch, whilst we counted our remaining rings and checked the notebooks. Our week's sustenance, John's department alas, for he was sadly neglectful of creature comforts, nestled in a kitbag – six sliced white loaves, which grew a splendid crop of mucor and penicillium within three days, and several tins of pemmican left over from his brother's Greenland expedition. Since Franklin the Arctic explorer, pemmican has been famous as a concentrated meaty protein, a hard brown wax which, when heated with water, provides a thick nourishing gravy. That was all we had. No fruit, no vegetables, no nothing.

I once camped in Glen Feshie with John and one day we climbed Carn Ban Mor to look for dotterel. Again it was blisteringly hot and half-way up he produced some dry ship's biscuits – hard tack – and nothing to drink. He was somewhat scrawny and his eyes bulged whilst the froth caked round his lips as he struggled to swallow this consolidated sawdust – like a tortoise forcing down a prickly pear. I believe he modelled himself on Thesiger. He would stride off into the hills alone and really soak up their atmosphere to the extent that, seemingly paradoxically, they produced a deep melancholy.

Had I never seen the guillemots of the Farne Islands, Ailsa's gannets, Rhum's shearwaters returning to their burrows on the high, dark night slopes of Hallival, sobbing so weirdly that the scalp prickled, I might have been content with the blackbirds of Wytham woods, near Oxford. It was there that I went in 1959, to join the Edward Grey Institute under the direction of that most erudite, yet simple and lucid scientific ornithologist, David Lack.

Wytham woods have produced more facts about great tits than anywhere else on earth. It is hard to realise just how much is now understood about this small, insectivorous bird, the oxeye, and, for some, equally difficult to understand why

4

it was worth discovering. At a conservative estimate something like a quarter of a million nests have been checked during the last thirty years or so. The whole exercise has been planned with the efficiency the military so rarely display – how many eggs are there in each clutch, what do they weigh, how fast do the young grow, how many successfully leave the nest, how many meals do they receive, how many die in the winter? On and on it goes. How many succeed in carving out a territory? Who mates with whom? How long do they live? In scientific terms, a deal more is known about great tits than about human beings. And it isn't only Wytham great tits that are investigated. Tawny owls, badgers, voles, woodmice, the caterpillars of the oak-eggar moth and the humble denizens of the leaf litter have repaid years of study. Nor is it a haphazard collection. At one time it was hoped that the Wytham food-chains – right through from the energy produced by photosynthesis and trapped by leaves to the sparrowhawks catching the great tits that ate the caterpillars – could be roughly quantified. It would have been an energy-flow budget of immense interest but, like comparably ambitious programmes in other ecosystems, it proved too complicated.

Put like this there is a sense of drama and excitement. Actually, finding out about great tits meant carrying a ladder through the woods to climb up to nest boxes, interminable trudging, scratched by brambles and bitten by insects. Chris Perrins spent many years doing this, until, as he says, he discovered that the branches were no longer as strong as they used to be.

In 1959 and 1960 our paths often crossed. I was beginning a three-year study of blackbirds and Wytham was part of my beat. It is a fine deciduous wood, full of bird song. There, in spring, the nightingale and the turtle dove still sang, the grasshopper warbler reeled secretively and the blackcap flashed his brilliant phrases. But my heart wasn't in it. In a sense I was trying to wear somebody else's clothes. David Snow had just completed a characteristically simple but elegant study of blackbirds in the Oxford Botanical Gardens and my remit was to extend this to farm and woodland blackbirds and to look for ecological differences between them.

The Edward Grey Institute and the Bureau of Animal Populations under Charles Elton were adjacent in the dignified old Botanic Gardens, to which came nuthatches and occasionally kingfishers (there was a pond!). David Lack dominated his Institute, and not merely cerebrally. The EGI was a family, tea-time a ritual. It took place precisely at 4.00 p.m. in the library, around a table, with Christine (the secretary) pouring and David presiding. His intellectual enthusiasm was infectious and amusing in its perfect catholicity. It mattered not whether the subject was drift-migration, egg-size, church history (a solo performance), babies' nappies or the quality of bread – his part of the discussion was structured, informative and above all, critical.

Across Oxford, in his collection of hen huts on the roof of the Zoology Department, Niko Tinbergen was developing an approach to animal behaviour that was already linking ethology to ecology, in a new way. His demonstrations of the adaptive value of behaviour based on rigorous fieldwork and experimentation, were having enormous impact.

This famous trio, Lack, Elton and Tinbergen, far from exhausted Oxford's talent. Henry Ford and Bernard Kettlewell were working on moth genetics, Arthur Cain on taxonomy and genetics and Pringle on the physiology of behaviour. Alister Hardy was still a force in marine biology. Oxford must have been one of the most stimulating centres for academic zoology on earth. Nor have I mentioned the lowly students of around that era – now not so lowly – such as Desmond Morris, David Snow, John Gibbs, Mike Cullen, Richard Dawkins, Hans Kruuk, David McFarland and many others.

So, with academic voices steadily blending in fortissimo and bird song on all sides in Wytham, why was I lying moodily amongst the brambles thinking of islands and seabirds? I like academia and I like woods and woodland birds, but when it comes down to daily living and going to bed contented, I think I had too much of the Fraser Darling romanticism. Islands, seabirds, escapism – that powerful blend – had got a grip. Here was I, ploughing through Wytham woods, virtually a suburb of Oxford, and returning to my bed-sitter to stew neck o' mutton and going to sleep with the roar of Banbury Road's traffic in my ears. Somewhere, there was a lonely patch which would hold a tent or a hut, from which I could watch the sea, bleak or sunlit as it may be. And there would be seabirds – gannets, fulmars,

6

guillemots, razorbills, puffins, kittiwakes, shags – raucous and elemental and vastly more exciting than blackbirds, dearly though I cleave to that melodious songster.

I tapped timidly on the Captain's door for my fateful chat. I wanted to desert the ship and join, not exactly a pirate crew but another outfit – Niko Tinbergen's research vessel. The tidy room bespoke the man. The writing table was devoid of all except the paper on which he was at that moment writing, the bookshelves empty except for the volumes germane to the current obsession. This was the cost-effective and organised way. Niko worked amidst a teeming profusion of books, papers, films, models, tanks and memorabilia, not to mention packing cases and field equipment. He, also, achieved results but perhaps at a greater cost to himself. In any case, my eminent Director was kind and helpful. He said that a researcher had to learn when to change direction. I had my new one tolerably well mapped. It lay about 450 miles to the north-east, in the outer reaches of the Firth of Forth, the 'emerald studded Firth', wherein lay the Craig 'callet the Bass', in Hector Boece's 1527 words 'full of admiration and wonder; therein, also, is great store of Soland geese . . . and nowhere else but in Ailsaie and this rocke'.

I still had to sell the idea to Niko Tinbergen, who hardly knew me, and I doubt if I could have done so without evidence that I could persevere with a behaviour study, alone, and get results. I had done my zoology at St Andrews and although it is undoubtedly the finest university in Britain (did not St Andrews' Bishops own the Bass Rock until 1316?) its Zoology Department happened to belong to Professor Callan. His consuming passion was the structure of the lampbrush chromosomes of the egg of the crested newt – by no means as esoteric as it may sound but equally not in the front line of behaviour. Being a delightfully small department, built around cell biology, neurophysiology and parasit-ology, there was no tuition in animal behaviour. But there were plenty of the good Professor's newts. As an ardent admirer of Konrad Lorenz I realised that newts court before they lay the eggs with which to further advances in the study of lampbrush chromosomes. The courtship of the crested newt therefore became the subject of my Honours Degree Thesis and, as though in preparation for the Bass, I had no option but to paddle my own canoe. Fittingly, for it was

7

Niko's creed, I prefaced my tome with a quote from the greatest of Scottish zoologists and a previous incumbent of the St Andrew's Chair, Professor D'Arcy Wentworth Thompson. In his classical *On Growth and Form* he wrote 'And all the while, like warp and woof, mechanism and teleology are woven together and we must not cleave to the one nor despise the other, for their union is rooted in the very nature of totality.' In his prose one catches the discipline of the great classicist and mathematician. Tinbergen's enduring theme has been the need to study both causation *and* function.

I laid my offering on the shrine and departed. It was enough. With the generosity and goodwill that have benefited so many, he agreed that instead of joining the team working on black-headed gulls at Ravenglass, I should hie off to the Bass and present myself each winter at Oxford for an intellectual clean-up. Fortunately there are gannets on the Bass for ten months, so my residence was almost continuous.

The lone worker can easily run out of steam and waste his opportunities but, even more importantly for a parent institution, he can damage public relations. Somebody owns the place in which he works and many people help with the operation and watch what is going on. Once soured, relationships can take a lot of mending, and there are plenty ways of souring them – scrounging, offensive habits, inconsiderate behaviour, and undue disturbance of the birds themselves. Then there is the safety aspect. So it was not altogether easy for Niko to let us loose on the Bass, even though Sir Hew Dalrymple, who owns it, was willing to let us live there. I can scarcely describe the elation that came with the sudden realisation that we had three whole years of island living, and seabird work, ahead of us; not a team – just us.

2

The Bass

And what would you have me to say
about a mass of homogeneous trap.

An island salt and bare
The haunt of seals and orcs and sea-mews clang.
      T. McCrie, *The Bass Rock*

In the Galapagos, to which we shall eventually sail, we were
well aware that we were living on young, volcanic rock.
Every step on the cindery slag made it plain, as did the fine
volcanic ash that settled on our tent, the fall-out from erup-
tions elsewhere in the archipelago. On the Bass, it was less
obvious but still the marks of its fiery birth were to be seen.
It rises from the seabed so clearly and strongly because it is a
hard rock, a compact clinkstone, the plug of a volcano. The
waves cannot wear it away as they have done the softer rocks
in which it is embedded. The cliffs are angular and sharp-
edged. Less than two miles across the Forth, on the Lothian
plain, they are much softer, fretted and worn . . . 'had the
Bass originally been composed of such a yielding tuff as that
on which the fortress of Tantallon is erected, we would now
in vain seek its place amid the waters.' There are other
clinkstone masses in the Lothian – Edinburgh Castle stands
on one. The name 'Bass' seems to be a descriptive term of
Celtic origin, meaning a hillock of conical shape and there
are, I understand, similar words in Norse and Icelandic.

The exposed stone of the Bass is a rusty, blackish brown,
distinctly red with iron in places, and streaked with sulphur.
Above the cliffs the naked rock outcrops particularly freely
on the south face, encrusted with green and yellow lichens.
The crude amphitheatre on the north-western slopes is presumably part of the original caldera and the loose stones and

9

boulders here and further round to the west are fragments of ejected rock. A kittiwake colony of some twenty-six pairs, now defunct, was built on what was the inner wall of the crater. Time has laid but a thin mantle of soil and vegetation on the boulder slopes and it takes only two or three years for areas newly colonised by gannets to change from green sward (often rather *too* luxuriant and green because of gull droppings) to bare earth with sharp protruding stones. A little later still, the underlying contours of the rock are revealed once again. Perhaps they have been covered and exposed many times over the millennia.

The Bass is imposing from all angles, looking more than its Ordnance Survey height of 350 feet. The cliffs, highest on the east, fall right away to the south and rise again, more round-shouldered, on the west. The open, south-facing slope falls away in rough tiers from the summit to the sea, first the Garrison garden, then the chapel, then the shelf on which now stands the lighthouse, built in 1902, and finally the rough, rocky apron on which the landing stages have been constructed. This, though, is partly man-hewn. Above the landings, the southern approaches are blocked by the ancient and formidable battlements which neatly plug the gap between the sheer east cliffs and the vertical south-west face. At right-angles to these inhospitable walls, a cannon embrasure overlooks the landings – at that range enough to daunt even the most foolhardy invader. Just how successfully this skilful blend of natural and man-made obstacles defeated intruders history tells, for when this ancient fortress was eventually dismantled in 1701 it had not been taken by force, though held by only a handful of men for years against the crown. Thus the well-known story of the four Jacobite prisoners, sent to the Bass in 1691, who overpowered the sentry and, joined by some dozen recruits, held out until 1694. Even then, it is said, they emerged not only with life and liberty, but with payment of arrears! No wonder Hector Boece described it as 'ane wounderful crag, risand within the sea, with so narrow and strait half [passage] that no schip nor boit may arrive bot allanerlie at ane part of it. This crag is callet the Bas; unwinnaibill by ingine of man'. Shortly afterwards, in 1706, Sir Hew Dalrymple acquired the Bass and Sir Hew still holds it, though renewed in the flesh through several generations.

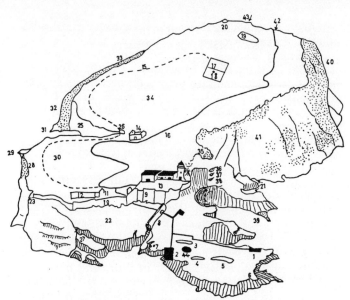

A sketch of the Bass Rock, looking north.

Key: 1 East landing (old N E port); 2 West landing (old S E port, actually on S side); 3, 4, 5 Brackish rock pools; 6 Osbairn point; 7 Site of old crane; 8 Mantle wall, with four cannon embrasures; 9 Battlement walls of old garrison; 10 Continuation of battlements; 11 Gable ends of old garrison living quarters; 12 Lighthouse outbuildings; 13 Lighthouse buildings and compound; 14 Fifteenth-century ruined chapel, on site of St Baldred's seventh-century cell; 15 Route followed by old garrison path; 16 Concrete path to N foghorn; 17 Upper walled garrison garden; 18 Well (drainage water); 19 Top loch; 20 Summit (ruined fortification); 21 Entrances to E caves; 22 S. slopes below battlements, covered with tree mallow; 23 W. turret; 24 E. turret, with lighthouse crane (unused); 25 N. side of cable gulley; 26 Gulley head pool; 27 Shag rock; 28 Headland colony; 29 The needle; 30 The headland; 31 Low shelf, a jutting outcrop at base of W S W face; 32 Nesting mass on S W face; 33 Nesting mass on W face, spreading beyond cliff-top; 34 Steep rocky face above chapel; 35 Lighthouse colony; 36, 37, 38 Upper E face, groups 3, 2 and 1; 39 Gannet colony at base of E face, ousting shags; 40 Nesting mass on E face; 41 Sparsely occupied sheer E face; 42 North foghorn colony, facing N; 43 Observation colony, facing N W; 44 Helicopter pad.

11

Actually there is a back entrance, up a steep vertical-sided gulley on the south-west, where a massive chunk of the south slope forms a headland partly separated from the western mass. But even the climb from sea-level to the base of the gulley is hazardous and then there is the narrow gulley itself to negotiate, no doubt staunchly defended at the top, in which case woe betide the climber. One boulder rolled down that chute would clear it in a trice.

This gulley marks the western entrance of a crooked sea-passage some 30–40 feet high which runs from a small, steep boulder beach on the west, right under the rock beneath the chapel, to emerge in a bisected, high-vaulted cavern on the east, alive with guillemots in season. Halfway through lies a slimy, boulder-strewn pool. It is a dark, unpleasant place, a 'dreary recess, full of chill airs and dropping damps'. Seals occasionally haul out in it and may even drop a pup now and then. The Bass defenders under siege used to receive food, and contraband wine from the French, and no doubt the valuables of the Earl of Buccleuch, and perhaps the Public Records of the Church of Scotland, which were sometimes taken to the Bass for safe keeping, were removed there when danger threatened. But all I ever saw were injured seabirds that had chosen this gruesome place in which to die.

I know the Bass, I've entered its dark bowels, I am thoroughly familiar with its seabirds but, sadly, my imagination is too ill-tutored to produce pictures of the innumerable human dramas which have been enacted on the 'auld craig'. All that flickers upon the inward screen is a kaleidoscope of disconnected fragments. What was St Baldred, saint of the Bass, like? What sort of life did he live in that primitive cell, hundreds of years before the fortress was built? He died on 6 March 606, '07 or '08 – some say 756, for apparently there were two Saint Baldreds and perhaps they were one and the same. Did he record anything about the gannets, which were surely there long before he was? Did he gaze across at the Lothians, straining to make sense of the skirmishing that went on as the Saxon King Ida established his dominance in the area, as elsewhere between the Forth and the Humber? And thousands of years earlier, what about the people of the Stone Age and the Bronze Age? Would they not have plundered this great store of seafowl? It is little more than a mile from the nearest point on the mainland and at slack water on

The Bass, showing the great east face.

a calm summer's day the merest cockleshell would get them safely across. A few men armed with clubs could reap a rich harvest, or wreak havoc, in a few hours. Prehistoric man doubtless devasted many of the more accessible seabird colonies around Britain. It is not even known when and by whom the garrison was built. Was it Kenneth, Malcolm II, or neither? It was in 1005 that Malcolm II conquered Kenneth III and later invaded the Lothians, which were subsequently ceded to him by the Saxons. What sort of life did the garrison soldiers live? What did the young Prince James, twelve years old, think of his month on the Bass in 1405, en route to France, though he ended up in prison for the next nineteen years? What did King James IV do when he visited the rock in 1479 after paying fourteen shillings 'to the botemen that brought the King Furth of the Bass'. Getting on for a century later, James VI was so taken by the place that in 1581 he tried to buy it for any price that Lauder might name, but Sir Robert refused. As a result of his visit the market crosses of

13

many towns north and south of the Forth heard open proclamation of a new order protecting the gannet and other seabirds from indiscriminate slaughter, under pains of heavy fines or imprisonment on the Bass. More than three hundred years later indiscriminate slaughter in the name of sport went quite unpunished! James was not the only monarch to covet the Bass. Charles I tried, and failed, to secure it. Ironically enough, when Charles II did succeed in laying hands on it, he used it for a grimmer purpose than any of his predecessors would have done.

And what about the traffic in gannets? Often, as I watched the season's crop of fat youngsters crowding the ledges and slopes, the contrast between their secure and protected state and the plundered ranks of years gone by crossed my mind. For hundreds of years the Bass gannets provided fertiliser, feathers, eggs, fish and meat, to the tune of hundreds of gold pieces per year. The return of the adults, fat from their winter feeding with nothing to deplete their energies except flying, fishing and keeping warm, was eagerly awaited. The early-season 'panics' were well known and the Garrison took pains to disturb them as little as possible. In Major's well-known words of nearly 500 years ago 'these geese, in the spring of every year, return from the south to the rock of the Bass in flocks, and for two or three days, during which the dwellers on the Rock are careful to make no disturbing noise, the birds fly around the rock.' The birds get to work repairing their nest drums, which had been removed by the ton for fertiliser. How well the Garrison garden looked in those days. Eventually came the crop of eggs. Surely the gannet's willingness to replace a lost egg, or even two, would have been well known – the St Kildans knew. And gannet eggs, although small for the size of the bird, are larger than a good hen's egg and highly palatable. They came from the cliffs by the basket-load, limy shells still streaked with dried blood, a sure sign of freshness.

The fish, doubtless mackerel and herring among them, were those which the gannets regurgitated, as they do when frightened. I have never, myself, seen a perfectly fresh fish vomited by a gannet, though some have had only part of the head digested. And even then, the fish have been warm and, as it were, partly cooked. This doesn't make them less edible but it may offend a squeamish stomach. It may well have

14

been fish procured from the honest gannet that the dishonest Governor of the Bass sold at inflated prices to the wretched covenanters, thirty-nine of whom were incarcerated between 1672 and 1686.

Probably something approaching a million gannets have been harvested from the Bass in the last thousand years. I don't care to dwell on the bludgeoning, and the maimed birds that fell into the sea to drift away – man is a messy predator. There was a certain grim justice in an accident which befell some of the hunters, though. As on St Kilda, it was the custom for the men in the gannetry to club their victims and toss the body over the cliffs to be collected from the sea by a boat crew. One such, unobserved by the business end of the operation, drifted close in beneath the beetling cliffs. To their consternation and terror, they were attacked by dead gannets, which came hurtling down to splatter around and into the boat. The surge of the sea and the wind drowned their yells and the fruits of their comrades' labour continued to make their final protest. A guga can weigh nine or ten pounds and the cliffs are high. The miserable men crouched in the rowing boat, trying to shelter beneath the splintered seats as the missiles rained about their ears and when the all-clear sounded they were tolerably incoherent, well-coated in gore and feathers and presumably satiated with gannets. But what can compare with this account from Stevenson's *Catriona*?

> At last the time came for Tam Dale to take young solans. This was a business he was well used wi', he had been a craigsman frae a laddie, and trustit nane but himsel'. So there was he hingin' by a line an speldering on the craig face, whaur it's hieest and steighest. Fower twenty lads were on the tap, hauldin' the line and mindin' for his signals. But whaur Tam hung there was naething but the craig, and the sea belaw, and the solans skirling and flying. It was a braw spring morn, and Tam whistled as he claught in the young geese. Mony's the time I've heard him tell of this experience, and aye the swat ran upon the man.
>
> It chanced, ye see, that Tam keeked up, and he was awaur of a muckle solan, and the solan pyking at the line. He thocht this by-ordinar and outside the creature's habits. He minded that ropes was unco saft things, and the solan's neb and the Bass Rock unco hard, and that twa hunner feet were raither mair than he would care te fa'.

'Shoo!' says Tam: 'Awa', bird! Shoo' awa wi' ye!' says he.

The solan keekit' down into Tam's face and there was something unco in the creature's ee. Just the ae keek it gied, and back to the rope. But now it wroucht and warstl't like a thing dementit. There niver was the solan made that wrought as that solan wrought; and it seemed te understand its employ brawly, birzing the saft rope between the neb of it and the crunckled jag o' stane.

There gaed a cauld stend o' fear into Tam's heart. 'This thing is nae bird' thinks he. His een turnt backwards in his heid and the day gaed black aboot him. 'If I get a dwam here' he thocht, 'its by wi Tam Dale'. And he signalled for the lads te pu' him up.

And it seemed the solan understood the signals. For nae sooner was the signal made than he let be the rope, spied his wings, squawked out loud, took a turn flying, and dashed draught at Tam Dale's een. Tam had a knife, he gart the cauld steel glitter. And it seemed the solan understood about knives, for nae suner did the steel glint in the sun than he gied the ae squawk, but haigher, like a body disappointit', and flegged aff about the roundness of the craig, and Tam saw him nae mair. And as soon as that thing was gane, Tam's heid drapt upon his shouther, and they pu'd him up like a tied corp, dadding on the craig.

The grease obtained from gannet fat was extremely valuable – Hector Boece's theology may be forgotten but his description of the properties of gannet fat deserves better.

> Within the bowellis of this geis, is ane fatness of singulaire medicine; for it helis mony infirmeties, speciallie sik as cumis be gut [such as come by gout] and cater [catarrh] disceding [diseasing] in the haunches or lethes [groins] of men and wemen.

The Bass gugas were no strangers to the Royal table, for as Gurney relates, among moorfowls, partridges, plovers, herons, cranes, dotterel, redshanks and larks, served on 4 September 1529 to the King (James V) at Edinburgh, were 'once solares', which are solan geese. Nor were they strangers to the rough Irish labourers who at harvest time each year came to the Lothian farms and feasted on gugas, roasted on huge fires at Cantry Bay, opposite the Bass.

In the late eighteenth century there were, apparently, climbers, who for part of the year, lived on the Bass in a little hut. Exactly where they placed it I don't know – perhaps where the lighthouse is. They sold liquor, bread and cheese

16

'for the accommodation of chance visitors and of the sports-men who frequented the place for the diversion of shooting'. Those were dark days for the gannets – shot for fun as they tended their young. Sometimes I am tempted to believe that we are improving.

Whilst England absorbed the Norman conqueror, tamed its barons, developed its judiciary and Parliament and be-came a world power the 'Auld Craig o' Bass' continued century after century thriving, toiling, warring and intrigu-ing. What a story it could tell. But perhaps its most infamous hour came with the incarceration of the Presbyterians who refused to accept King Charles II's authority in religious matters and especially rejected the government of the church by bishops, which Charles attempted to force upon them. What a blood-bath it was. Indictment, trial and execution all in one day and on false evidence. With every gaol and tol-booth crowded, where better than the Bass, near to Edin-burgh and impregnable, for imprisoning leading covenan-ters? So they came, and endured the cold, damp, starvation and solitude. I know what it is like, sleeping in winter in those unheated ruins and how they withstood it for years, inadequately fed and clothed and far from young, defeats me. I have spent many a night close by the spot where Blackadder died:

> So the bless'd John, on yonder rock confined, –
> His body suffered but no chains could bind
> His heaven-aspiring soul; . . .

The great Charles, cultured patron of the sciences and foun-der of the Royal Society, the most prestigious scientific so-ciety in the world, bought the Bass specifically to imprison upright and conscientious men after reneging on his Solemn League and Covenant to maintain the privileges of the Church of Scotland.

In 1706 the Bass passed into the hands of the Dalrymples, who still own it and who care for it lovingly. A quick skate over the intervening two-and-threequarter centuries would note that the Bass has seen the end of gannet culling (around the 1880s), the building of the lighthouse (1902), the closure of the Bass and the mining of the Forth during the Second World War, the demise of the stately Lighthouse Relief vessels with the arrival of the helicopter pad in 1974, and the steady increase of its most famous inhabitants, the gannets,

17

to their present total of some twenty thousand occupied sites. Another page will be written when the light becomes automatic, as it is scheduled to do. Then, once again, the Rock will stand without a human soul:

A distant giant range, . . .
North Berwick Law and Bass amid the waters.

3

Domestic

The baffled wind gathers . . . and comes down with
a terrific buffet. The hut shakes visibly, and tired
though we are and falling asleep between whiles,
these dunts of tortured air wake us suddenly again
every time.

F. Fraser Darling, *A Naturalist on Rona*

We ferried 2,238 pounds of concrete to the Bass – more than
fifty large concrete cones to be cemented into the gannetry as
markers, every one of which I had to land from a small boat
and lug up the Rock in a rucksack. Some are still there,
nearly a quarter of a century later, nor have I yet thought of
any better way of marking solid rock amidst nesting debris
with conspicuous, individually recognisable symbols. But
today, 26 February 1961, the load was merely a Land Rover
full of 'junk', including arsenical soap, pot eggs and cast-iron
markers. Alongside North Berwick harbour, after a frugal
breakfast in Fred's store,* redolent with creosote and old
lobster-pot bait, we set out for the Bass in the old *Norah,* a
double-ended 'Fifer' of some twenty feet overall. Already
the air rang with the alarm calls of the herring gulls and the
gannets had been back for weeks. On a fine day for once, the
inner landing was quiet enough. The half-tide exposed the
barnacles and limpets but the shiny brown straps of the
laminaria were still well down. How many dragging feet had
passed that very way through the centuries. Sir Walter Stew-
art on his way to the executioner's block at Stirling, reluctant
soldiers, dejected covenanters facing incarceration in the
Bass dungeons and even modern lighthouse keepers, ill-

* Fred Marr, our old and valued friend, is the official boat-
man for the Bass Rock.

pleased, sometimes, to leave home and hearth for the winter comforts of the Bass Rock. But my feet were willing enough. The following day was a beauty. All over Britain the temperature rose 10–20 degrees F above normal. But the sea, still only 45°F, gripped like a vice. I could well believe that survival time in the North Sea was only three or four minutes. A fortnight earlier had been the hottest February day of this century – 65°F on the Air Ministry roof. I had thought to try out my old camera in the park in Oxford but the many amorous couples had thwarted that idea. Birds, too, were in full song and a pair of blackbirds had a nest with eight-day-old chicks. A year later, to the day, I lay out on the summit of the rock in a north-easterly gale attempting to identify colour-ringed gannets. Showers of snow and hail swept over, filling my ears and binoculars. This extreme variability between years is one very good reason why most British seabirds either lay their eggs over quite a period or, in a few cases, not until rather late in the season. Bad weather may not affect adults directly, but it can disrupt their food supply and kill their young.

These wretched moments – and there are many – disappear, never to emerge in scientific papers. The gannets just sat there, headless lumps sleeping snugly, sublimely indifferent to the elements. It must be lovely to sit on a cliff ledge in an arctic blizzard, as warm as toast and as comfortable as a cat on a cushion. Clouds of spray hurtled over the east landing. Thousands of tons of seawater foamed onto the unyielding basalt. Even as it separated into a myriad coursing rivulets the wind whipped it off the glistening rock. The bulking east cliffs gleamed dully and the gannets, refusing to vacate their ledges, were jostled and lifted bodily by the screaming wind. The sea temperature had fallen to 40°F.

Six days earlier, it had been mild, sunny and calm. Thousands of mosquitoes swarmed by the garrison garden. A swallow could have eaten its fill in a few minutes. An early bee was out and about. Such weather is often followed by mist and the next day cotton wool had enveloped the sea but left the upper air clear, so that the top of Berwick Law loomed preposterously high.

Sleeping on the Bass in the depths of winter was a cold job. Despite layers of newspaper, we always awoke with the side nearest the floorboards like a slab of frozen mutton.

But I'm racing ahead.

I first set foot on the Bass on 18 July 1960, and lodged in the lighthouse. Today, though not for much longer, the tower is the same and the paraffin-powered light still operates, but my old-fashioned living quarters have vanished. In the old days the living room, lined with varnished pine cladding, held a brightly burning coal range and spotless oven. A large, rectangular copper, lovingly burnished, sat to one side and provided scalding hot water the colour of strong urine, 'a dilute tincture of guano'. Coming from the seepage well near the summit of the rock, it derived its yellow hue from the rock and humus through which it percolated. A century and a half ago the same well was described as 'full to overflowing with a brown, turbid fluid . . . which had proved the grave of a hapless sheep during one of the snowstorms of winter'. The prisoners of the Bass sometimes had to drink this 'corrupted' water, sprinkled with a little oatmeal. The lavatory, somewhat confusingly flushed with the same solution, was outside, round the back, far enough in a January gale. Baths, if any, were taken in a large, shallow 'pan' placed in front of the fire – just as my mother used to bath us. Commodious pine cupboards ('presses') lined one wall and an old wooden clock with Roman numerals ticked on another. The lamp, a pressurised Tilley hanging from a hook in the ceiling, hissed companionably as it shed its light on the plain wooden table, where so many letters home were penned in the lonely watches of the night. The sleeping quarters opened directly off the living room, each narrow cell with two-tier bunks in solid old pine. In the evenings, a black cat in its seventeenth year, with a kink in its tail, slept by the fire whilst the keepers read, chatted, smoked and drank tea. Now, the coal range lies at the bottom of the sea; the heating is electric, the keepers enjoy a modern bathroom, a kitchen, colour television and separate, single bedrooms. Bright, synthetic materials bloom everywhere and there is no cat.

We were fortunate in that, throughout the years in which we lived on the Bass, the same four keepers, George, Sam, Bert and Donald, remained there. They were all of the old school, from traditional keeper families, or ex-seamen, and with years of service behind them. Nowadays recruits are harder to come by, despite vastly improved pay and conditions, and fewer of them stay long. It seems odd, but I suppose expectations are greater and fortitude rarer. But the trend towards automatic lights is a retrograde one, for the keepers do far more than merely keep the light burning. No automatic system can possibly provide the safeguards and information that they do. And if the service is dismantled it may prove difficult indeed to resuscitate it.

But these things are superficial. The Bass remains unspoilt, unexploited, as full of history as of birds and arguably the finest seabird colony in Britain. It changes its moods season by season as the seabirds return, filling the cliffs with noise and motion, and depart, leaving them empty and desolate in winter.

In 1960 I had to get to know the Bass, choose groups of gannets for detailed study, colour-ring some of them, prepare for permanent living on the Rock, starting early the next year, and put up observation hides at my main colony. One summer visitor, a truculent Yorkshireman, took me soundly to task over this. He said, in that bluff, call-a-spade-a-spade voice, that it was sheer pretension – nesting gannets were so tame you could walk right up to them – he had often done so. Of course you can, but you will not see much truly representative behaviour. I daresay he did not know the difference between a relaxed, a suspicious and a downright scared gannet. Disturbance is in fact one of the great vitiators of behavioural and ecological work. I was myself at first unaware just how important the human observer is in influencing the course of events in his study colony, but he really can have a serious effect. For example, he greatly influences predation. No gannet, except occasionally a young female breeding for the first time, loses its egg or chick to a herring gull unless it is disturbed by man, in which case its attention is diverted and its reactions directed towards the human intruder. First-time breeders sometimes desert their eggs soon after laying and a gull may then take them. But normally, gulls do not dare to approach close enough to steal an egg. In some seabirds, guillemots for example, or terns, the effect of disturbance can be catastrophic. Even if there is no outright loss, disturbance can retard laying. And the effect of disturbance on behaviour is no less forceful. Darling, of course, was largely immune from outside disturbances when studying his herring gulls but the Bass is much more vulnerable, and when visitors were on the Rock I made it my

22

business to sit in my hide and deflect unwelcome intrusion, otherwise my results would have been worthless.

Like Darling and Ronald Lockley on their little islands, we hoped to make the Bass Rock a satisfying home for a few years. I have been on Lockley's Skokkholm when the spring-time bracken and bluebells were out and the Bass is harsh by comparison. It is, after all, only seven acres, much of it bare rock. There was only one suitable spot to live – a few square yards of flat ground, south facing, about 150 feet above the sea and looking towards Tantallon Castle on the Lothian plain. It had been chosen by others before us, notably in the seventh century by St Baldred. The Bass chapel, consecrated in 1492, was built on the same spot and its ancient, storm-fretted walls still remain. Within this shell we placed our modest little cedarwood hut, twelve feet long and eight wide. We were to cross that venerable chapel threshold more than 30,000 times in the next three years. Who, on a sunny morn-ing such as those seen by the ordinary visitor, can imagine it on a tempestuous winter's night? 'What time the midnight moon looks out through rock and spray and the shadow of the old chapel falls deep and black athwart the sward'.

That hut may have been modest, but each section had to be transported to the Bass, lugged from the boat to the landing and carried 150 feet up the rock. It came across in the old *Norah*, who was not beamy enough to take the pieces without letting them overlap the sides. She came in to the inner landing to take advantage of the hand-cranked crane but there was too much swell, so she motored round to the outer. The rolling motion caused the underlying sections to slide out from beneath the upper ones and Fred's father, then over seventy, skipped nimbly from side to side in a vain attempt to halt the oscillation. With a final tilt, one section slid overboard, ripping the seat out of the old man's trousers on a protruding nail. The rest of the hut followed, and then the furniture, bobbing slowly down the Firth on the ebb tide.

But landing on the Bass is not as tricky as many early chroniclers maintain, though the approach to the east (outer) landing, beneath the towering cliffs and past the dark caves, alive with gargling guillemots and sepulchrally croaking shags, is as fine an introduction to a seabird rock as one could wish. After the excitement, I wandered down to the landing in the fading light. The shags lined up in scores. A seal

23

examined me from a distance – a black ball on a grey sea. Gulls wailed and a pair of purple sandpipers pottered over the exposed seaweed. This was our garden. In a few crowded hours the visitor, surrounded by others, cannot begin to capture this feeling, that one belongs.

We might have had to wait weeks for a day calm enough to carry the hut sections up to the chapel, for they catch the wind like a sail. But the next day was flat calm and once I'd cleared centuries of nettles and tree mallows from the inside of the chapel George, the Principal Keeper, laid the floor on stone piles, erected the walls and put the roof on. The total building cost, carefully itemised on the back of an old envelope, came to £3–9s–9d. We passed stout hawsers over the roof and secured them to eye bolts driven deeply into the base of the chapel walls. Nor was this precaution, which left to my own devices I probably would have ignored, superfluous. Even within the sheltering walls the wind, funnelled up the gulley to the south-west, lifted the roof till the wires groaned, and bulged the walls so vigorously that crocks fell off the shelves.

Inside the hut the sweet scent of cedarwood, especially fragrant on warm days when the aromatic oils volatilised, remained throughout its life. Whenever I smell cedar I remember the hut, with its cheerful red and white checked curtains and all its companionable clutter. The total cost, including transport, building materials and all contents (except the fridge) was well under £200. The lighthouse reliefs came every fortnight, so perhaps we needed the fridge although generations of keepers had managed without one. Their sturdy and commodious old whitewashed safe still hung inside the lighthouse tower, alongside the freshwater tanks, but it didn't preserve meat and butter for two weeks in summer. The safari camp beds came from Milletts and the wooden chairs cost 18/9d (93p) each from Woolworth's. Our hot water came from an old copper urn sitting on top of the flat-wick paraffin oven which cooked many a batch of fine Yorkshire puddings. We carried the water down to the hut from the old well near the summit of the rock, in a plastic dustbin – a most unyielding chore. We bathed in the sea or else in the washing-up bowl and buried sewage in holes laboriously picked out of the rocky slope above the chapel. Like the Queen, we ran a tight little ship.

In the hut.

I have always been as independent as possible. Nowadays it is usual for a field scientist to look to his Institution or Government Department for everything from equipment, including binoculars, telescopes and cameras as well as the more technical items, through personal gear such as sleeping bags and protective clothing, to out-of-town living expenses. Throughout twenty-five years of fieldwork I have usually used my own equipment and personal gear and I have never claimed a single day's subsistence. I would be interested to know how much money could be saved by the Scientific Civil Service if they made such economies where practicable. I can just imagine it!

Despite this, we were glad enough to accept the help of Her Majesty's Commissioners in getting our food across to the Rock. Fortnightly reliefs were red-letter days. There was a marvellous store in Leith, Coopers by name, which made up the keepers' orders, and ours, sent ashore with the off-duty man. Then, at the next relief, out came the goods. Somebody in Cooper's was a saint among shopkeepers. Every item, listed and priced in copperplate handwriting, was painstakingly wrapped in newspaper or tissue and expertly packed. Year in, year out, there came to the Bass, as

25

to the other stations of the Forth, those heavy, green-painted wooden boxes with rope handles and rope lashings, full of good things. I used to imagine this anonymous perfectionist, a stooping, brown-coated figure, bespectacled, unhurried, working reverently in an old-fashioned shop smelling of coffee and cheese, wrapping and listing, wrapping and packing, wrapping, wrapping. We loved relief day, with its fresh meat, fruit, vegetables, cream and new bread. Often enough I returned to our little hut, from a session 'up the hill', to find a flushed and dishevelled wife in the midst of pies and scones (usually cooling on the floor), vegetables and potatoes bubbling on the stove, chops frying, paraffin oven going full blast and the whole blending deliciously with the scent of cedarwood. Then we would sit down to a feast. The tail-end of the fortnight was a different story.

One of my favourite suppers was rabbit pie. There are no rabbits on the Bass now, but until about 1958 they were abundant, replacing the twenty-odd sheep which a Castleton farmer used to graze there. Bass mutton was famous. The eastern and south-facing slopes supported a fine turf of *Festuca* of a lovely glaucous variety, which has been largely replaced by the soft, fleshy, grassy luxuriance of *Holcus mollis* 'Yorkshire Fog', not nearly so attractive. The turf tussocks are dissected by deep runnels and the cover looks as though it would suit puffins, but old gaudy neb never penetrates beyond the Garrison walls, within whose ancient crevices about fifty pairs nest. They give the impression of a small colony, hanging on in an unusual but traditional nesting place but there are large numbers elsewhere in the Forth, on Craigleith and the Isle of May, and doubtless there are immigrants to the Forth islands. The spectacular increase on the Isle of May, documented by Mike Harris, from 5–10 pairs in the early 1950s to an estimated 10,000 pairs in 1982 shows that this does happen and in fact some immigrants are known to have come from the Farnes. I picked up a puffin on the Bass that had been ringed on Grassholm in the Bristol Channel which is not only far to the south but on the other side of Britain. It may be that the Bass is less suitable than it seems, or, perhaps more likely, that there is no real pressure on nest sites and new breeders are attracted to the larger, more conspicuous colonies. For a highly gregarious bird like the puffin there is obviously some advantage in numbers. This

26

The lighthouse relief.

subject is one of the least explored aspects of seabird biology. Nobody in the world can tell you why each species nests at the density it does.

We could not catch our own rabbits, but there was no shortage of eggs. Gull's eggs and chips was one of my favourite meals. Like the keepers, we used to preserve scores of eggs in waterglass and they made excellent cakes. Herring gulls are persistent layers and continued to replace their lost eggs until the pigment glands failed and the eggs were pale blue instead of green or brown, blotched with black. I have always strongly disliked robbing birds' nest in any shape or form, but in this case the eggs would have been collected anyway, and the gulls were undoubtedly a nuisance to the other seabirds. It requires no invidious comparison to realise that a Bass Rock teeming with herring gulls and losing some of its other seabirds would be impoverished. The conservation of vulnerable species does sometimes require that others be controlled. It cannot rightly be argued, nowadays, that nature should be left to her own devices for we have already had a good shot at wrecking the machinery. Herring gulls, for example, would probably not have become the pest they now are without the extra food provided by man.

The old walled garden of the Garrison, near the summit of the rock, was (in the 'sixties) tended devotedly by the keepers. Part of it grew tatties for the lighthouse, and wonderfully buttery they were, while the remainder was divided up between the keepers and its produce went ashore along with whatever they had made in the previous spell – perhaps a small table, a jewelry box, or in George's case a superbly crafted writing bureau, well wrapped up in old sacking. Significantly, since the coming of television all this activity has ceased, the deadening hand of the box, writ small. As incomers, we cleared a small patch outside the wall and another right on top of the rock, looking north to the Isle of May, nine miles away. We planted Midlothian earlies but the yield was slight, and everything generously fouled by the herring and lesser black-backed gulls, though if we had humped enough compacted gannet muck over from the edge of the east cliffs we could have done better. It makes excellent fertiliser, rich in seaweed and guano and, unlike the seaweed in Galloway where I now live, unpolluted by nuclear waste. Outside the garden dense clusters of small, wild daffodils flowered each spring, as they must have done for hundreds of years, near to a gnarled old elder, beloved of dunnocks.

Miller writes: 'the garden, surrounded by a ruinous wall, and a broad fringe of nettles, when seen in the genial month of June, 1842, bore, among the long, rank grass . . . its

28

delicate sprinkling of garden flowers grown wild.' The cherry trees, which it once contained, have vanished, but the elder tree survives. On a bright, calm spring day – miserably rare – with singing rock pipits parachuting to earth, early willow warblers flitting along the lichened walls, gulls above and the clamour of gannets all around, it is a blissful spot 'full of noises, sounds and wild airs, that give delight and hurt not'. Skeins of gannets come in from the north, past the May, and one day, I don't doubt, they will recolonise her cliffs.

These early spring days had a great feeling about them. A fresh field season was under way with its new opportunities. The rock was shaking off winter's dead hand, and the Bass, fast-frozen and snow-powdered in a January gale, leaves ample room for improvement. When the migrants began to appear, anything could turn up. Throughout winter, snow buntings often foraged among the frozen gannet drums. From mid-March, large 'falls' of robins, thrushes, black-birds, redwings, bramblings, goldcrests and hedge-sparrows could follow mist and north-east winds. A little later came falls of wheatears, redstarts, willow warblers, ring-ousels and others. In fact, we had a chance of attracting almost anything that the Isle of May does, and that means most birds on the British list. Some exciting raptors came with the passerines. A peregrine (these marvellous falcons used to breed on the Bass) came to grief, presumably because it failed to pull out of a stoop. We found it jammed headfirst into a crack in the rock. A merlin was seen in hot pursuit of a house sparrow, which it followed down to the battlements in a headlong dive before securing it and carrying it back to its perch above the chapel, where it plucked and ate it. Years later, in Jordan, I saw a merlin casually snatch a red-rumped swallow in full flight, with one foot. Sadly, merlins are disap-pearing from many of their remaining strongholds in Britain. As a youth I found their nests on the drear, millstone grit moors of the Yorkshire/Lancashire borders. In Galloway afforestation of the moors has swallowed up merlin habitat. But this is not the whole story, for there has been a shift in the distribution of merlins. They are pulling out of their more southerly haunts and pushing further north.

The first waves in the spring migration of common gulls to the north-west was an event so reliable that one could have set a calendar by it. On 18 April thousands passed high overhead, way above the clouds, their plaintive, querulous calls floating down. It happened within a day of this date in three successive years.

Then would come the wind. Spring gales were less trying than summer ones, because they were part and parcel of spring, as also of autumn. They were not liked, but they were expected. June blizzards of hail and sleet were definitely below the belt. So were gales or near-gales that persisted for eight days non-stop, especially when the radio insisted that we were having fine weather. Islands are so much colder and windier than inland areas and I hated those westerlies. They soughed and roared without respite, buffeting us off balance, mocking any attempt to use binoculars or write in note-books; eye-watering, nose-dripping irritants. Often the hide, securely guyed and weighted with boulders, was blown clean across the rock – fortunately never out to sea and never with us in it. At night the wind exploded against the hut, twanging the wires and threatening to lift the roof. In the blackness, with the noise of the wind-lashed sea, I thought of the single-handed sailors who go down into the roaring forties and the tempestuous fifties. Imagine being alone in a small boat, in the almost continuous darkness of a sub-antarctic gale, amidst those awful seas that have engulfed many a fully-crewed schooner. Maybe they think they will be lucky with the weather and by the time they are in it they have no choice.

The gannets find it difficult to land in a gale, pitching awkwardly onto their breasts in a flurry of balancing wings. Damage caused by awkward landings is the main natural cause of death in adult gannets. Unlike some British sea-birds, they seem virtually immune to starvation. This is hardly surprising since they can survive for two or three weeks with little, or no food, and then replenish their re-serves when they locate a shoal.

We had our own swimming pool on the Bass. At the top of Cable Gulley on the south-west, just where the path bends round towards the chapel, a deep hollow has been blasted out of the solid rock. Possibly the lighthouse builders did it, so that seepage water could collect. Then they could use it for mixing hundreds of tons of concrete for the path. By 1961 it had filled with rocks and mud and was covered over with grass and nettles. How blithely I began. Had I but known, it was more than five feet deep, with sheer rock walls. Every

30

lunge of the spade hit a stone and every bucketful of stones and mud had to be squelched out through the clinging, stinking mud, and slung down the gulley. It was a real labour but when eventually it had been cleaned out it began to fill with vile, deep yellow seepage water. Had you chanced unexpectedly upon the scene, on many a summer's day that followed, you might have wondered at the sight of Adam and Eve leaping into this sinister hole.

The Bass, like the Farne Islands and the Isle of May, is popular with summer visitors, most of whom treat it with respect. It would be hypocritical to pretend that we welcomed the boatloads. Few fieldworkers are pleased to have their routine upset and for me it could mean hours in the hide, tensing every time the gulls set up a cacophany of alarm calls, popping my head out to deflect the bird watchers who, carelessly or deliberately, were about to wander into the fringes of my study group. This would have opened the way for egg-stealing herring gulls and many of my nests would have lost their contents. But what is the point of writing graphically about nature, and shooting marvellous films for television, if the interest engendered cannot lead anywhere? It would be quite wrong to deny people a thrilling experience which, moreover, is likely to make them more than ever concerned to protect wildlife. The thousands of people who set foot on the Bass take a memorable experience away with them. Only a few are bloody-minded and a tiny minority leave their beer-cans, plastic wrappers and film packets lying around. The photographers and bird-ringers caused the most disturbance. Sometimes the rock was alive with crawling, creeping forms, dangling cameras, meters, gadget bags and tripods and oozing concentration. Terse communications flew between them: 'Used all my Ferraniacolor, Bill', 'Think I'll try the long lens, Alec', 'Did y' bring a changing bag, Dick?' Once I came over the Rock and saw a black-coated figure slap in the middle of one of my special, colour-ringed groups, waving his arms like a dervish. I simply couldn't believe that anyone could be so daft. It was a sweet old Reverend, who merely wanted to photograph gannets flying back in to their chicks. Apparently he didn't notice that these same chicks were about to attempt the impossible feat of supporting their fat little bodies in mid-air, with no wings.

Over the centuries, the Bass has received many distinguished visitors. Harvey, the physician who discovered the circulation of the blood, accompanied Charles I to the Bass in 1641 but can his word be trusted when he says that the surface of the island was almost completely covered with nests 'so that you can scarcely find free footing anywhere'? Possibly, though, he merely sailed round it, without landing. Other early chroniclers, such as Slezer, Sibbald and Mackay, said much the same. Indeed, in 1831, Macgillivray claimed that nests were placed on all parts of the rock, being most numerous towards the summit. He implied, also, that they nested on the slope below the battlements, where nowadays they certainly do not.

The famous naturalist, John Ray, visited it with Willughby and Skippon in August 1661, from Dunbar:

> The young ones are esteemed a choice dish in Scotland, and sold very dear (1/8d plucked). The beak is sharp pointed, the mouth very wide and large, the tongue very small, the eyes great, the foot hath four toes webbed together. It feeds upon mackerel and herring, and the flesh of the young one smells and tastes strong of those fish. The other birds which nestle in the Basse are these: the scout [razorbill], which is double ribbed; the cattiwake [kittiwake], cormorant, the scart [shag] and a bird called the turtle dove, whole-footed, and the feet red [black guillemot].

The black guillemot certainly no longer nests on the Bass, if it ever did.

In our day we enjoyed visits by James Fisher, Roger Tory Peterson, Niko Tinbergen, Konrad Lorenz and Gerard Barends, among others. Niko Tinbergen and I made a film about gannet behaviour which we called 'Gannet City'. He stayed with us in 1961 and 1962 and a more delightful visitor never stepped ashore. He insisted on helping with the fieldwork, which included weighing young shags on June's kitchen scales (I had lost my spring balance and was waiting for another one). He told us how he came to settle in England, at Oxford. Before that he held a chair at Leiden University. When he resigned to come to Oxford, a move which involved loss both of status and salary, the University Senate would not believe that he was doing it merely in order to start ethology in England, and offered to increase his salary. When he came to Oxford his annual laboratory budget was £25. Then Fords and The Nuffield Foundation made awards,

32

followed by the American Air Force and our own Nature Conservancy. Finally, at the time of his visit in April 1962 he and a colleague had just been awarded £50,000, by America, for work on behaviour in relation to electrical stimulation of the brain. It is perhaps worth noting that this massively funded project (£50,000 in 1962 is nowadays worth perhaps £200,000), yielded rather little of fundamental value compared with much of his shoestring work of earlier days.

I have never forgotten his stories of life in a Nazi Concentration Camp in Holland. Imagine what it must have been like, lying on a plain wooden bunk in a dormitory when the door was kicked open in the middle of the night and jack-booted Germans tramped down the hut, stopping quite arbitrarily by somebody's bunk. That somebody was taken out and shot in reprisal for a German killed by the Dutch resistance. Niko related this in a quiet, matter-of-fact voice but it has remained, for me, a classical horror story. The element of dread and anticipation so preyed upon the minds of some of his compatriots, distinguished men, that they reverted to infantile behaviour, utterly demoralised.

Some of his stories were less grim, like the family's eating utensil nicknamed 'the embryo fork', which was used to strain foetuses from the part-incubated gulls eggs which they were sometimes able to gather for eating.

Soon after Niko left, the author who first fired my enthusiasm for birds, now more than forty years ago, came to the Bass. At school, beneath my desk lid, I used to lose myself in stories by Kenneth Richmond – *Krark the Carrion Crow*; *The Heron Garth*, and *Kestrel Klee*. Listen to this on the hatching of Garth:

> How livid, unlovely he was – a rebellious gargoyle; and yet oddly beautiful with that splash of colour at his throat where the skin hung in a yellow pouch. His bill was short, enlarged at the end as though swollen after its long battering. His whole aspect was wry and wrinkled. His legs were blueish, pulpy, the toes enormous. He was born, in fact, splather-footed.

Then, as Krark the old crow was surprised at the heron's nest and speared by the adult heron,

> the old crow had almost ceased to be recognisable. Cleft through the spine, his wings fallen apart, only a shabby scarecrow was left to swing in the branches. A shredded husk

33

that twirled on its gibbet, a few worn feathers that settled slowly, lingering upon the air that had so long sustained them.

Heady stuff. I was caught at just the right age. Richmond himself I found brusque, not particularly articulate and certainly no scientist, but I owe him an enormous debt. And he gave me a piece of excellent advice. 'When you write your gannet book', he said 'for God's sake make it interesting. It is too fine a bird for mere facts and figures.'

Oddly, we had hardly any time for reading. Fieldwork of one kind or another filled the daylight hours, except during vile weather. Then came the chores. Finally, the ever-mounting pile of notebooks filled with my untidy scrawl, demanded more and more time as the information within them began its slow and toilsome progress from the original observation to the integrated item in a thesis or scientific paper.

My diary tells me that we loved the simplicity of life on the Bass Rock. If we did, it is a lesson that we dismally failed to learn, for ever since our lives have become steadily more complicated and our pile of non-essential possessions ever larger. But I know what I meant. There was a simple rhythm to daily life, a small circle of faces, a sense of remoteness from the distractions which consume time and leave little to show for it and the chance to identify deeply with one small and enthralling spot. I am sure I was wrong when I wrote, as I did, that I could happily live for ever on the Bass. But it shows how rewarding such a life can be, when a young man can even entertain such a thought. And June was scarcely less enthusiastic. Maybe it is essential to have a sufficiently important focus. Just to be on the Bass would not have been nearly enough. To be there for a purpose added the necessary ingredient. And if one is honest, how can the material fruits, which success in the endeavour would be likely to bring, be left out of the calculation? At this point, as always when personal motivation is considered, things become too complicated for objective analysis.

I suspect that Fraser Darling was equally ambivalent, for he eventually progressed far from the simple life of Rona, Tanera, and the Treshnish Isles. But in his earlier days he was marvellously able to enthuse his readers with the delights of island living, the great outdoors and his own intuitive, naturalistic kind of research. He had a focus, and he approached field work romantically. To him, it really mattered that the boggy patch near the tent should be left undrained because a snipe regularly fed there. He would have been troubled if the bird had flown in, as usual, only to find its little feeding place gone. That, to me, is a marvellous tribute. It wasn't that his horizons were so narrow that a trifle loomed large, but that the source of his vision was right there, with the snipe. The muse was in him and the careful garnering of hard data came less easily than it might. He knew he had to do it and he was interested enough in his ideas to look for evidence. But his approach was never primarily that of the hard-nosed scientist. It was the joy of living on Rona, as much as the chance to study seals and seabirds, that took him there. Maybe I have a foot in both camps.

Fraser Darling was an ecologist and not an ethologist. Although he wrote quite a bit about behaviour it was not really good stuff. Perhaps he suffered from a lack of formal training. But he was certainly a synthetic as against an analytical thinker. If his wider sweep, which the synthetic thinker naturally develops, passed beyond the boundaries of what we narrowly regard as science, it merely gave his writing an extra dimension. And a rare one. For every Fraser Darling there are ten or a hundred more conventional scientists.

4

Getting down to it

> The smell on the Bass Rock arising from the birds is
> not altogether pleasant, indeed a man who is not
> accustomed to it may be almost overcome . . . This
> obnoxious odour is augmented by the liquid arising
> from rain and spray mixed with the excreta of the
> birds and the remains of fish.
>
> J. H. Gurney, *The Gannet, a bird with a history*

Field ecology and behaviour demand protracted, routine
observations and can be pretty dull. Sometimes the problem
is simply how to do the work. How do you count a teeming
seabird colony, or even worse, a population of field-voles or
weasels? Occasionally a clever technical device provides information that would otherwise have been virtually unobtainable. Tom Royama's automatic recorder at the nest box
timed and photographed each visit of the parent great tit
with food. Usually, the prey could be identified from the
picture. Hilary Fry constructed artificial bee-eater burrows –
a bit like hospital urine bottles – which he inserted into the
bank where the birds nested. Inside each of them was a
sensitive weight recorder so that the weight of each meal
brought in, and each faecal pellet taken out, could be auto-
matically monitored. It is amazing how many fundamental
questions depend for their answering on details such as these.
Indeed, the question is itself the important thing. Niko
Tinbergen was superlatively good at rigorously defining precisely what he was attempting to do and why. Only then did
he think about how to do it. His famous study of egg-shell
removal in the black-headed gull was a case in point. It was
one instance of behaviour with a function that could be
analysed experimentally. Why do adult black-headed gulls

remove their egg shells from the nest area after the chicks have hatched? The evidence suggests that by so doing they reduce the chances of their young being spotted and taken by bird predators. The jagged white rim of the broken shell is a give-away. But to turn this from an idea to something more concrete (though less concrete than a fact) meant adducing evidence. Are nests with shells nearby more likely to attract attention? This could be tested. Even the observation (and this *was* a fact) that shells were not removed immediately after hatching was potentially significant and led to the suggestion that removal was delayed until the chicks had dried out, in which fluffy state they were less easily swallowed by a predator and therefore less quickly available.

These aspects of fieldwork – the hammering out of the problem and the ways of tackling it – impinge with particular force on the loner. Time can pass pleasantly enough on your island and it may be only later, when you try to write a coherent account of your findings, that the awful holes in the fieldwork become apparent. The problem is less acute if, either as a member of a team or by inclination, you want to find a nice, tidy little problem, but as I have said, Fraser Darling did not work like that and nor do I. A tidy little problem might, for example, ask whether breeding carrion crows that are given extra food within their territories rear more young, or rear young faster than others. The results should show one thing or the other. Similarly, I could have concentrated on the biology of the growth of the young gannet and planned all my observations and experiments around that. But studying simply 'the gannet' means following a dozen lines at once. This can easily lead to superficial and unstructured work. I found it impossible to quarry anything of value from the account of hundreds of hours of gannet watching, at Noss, carried out by Richard Perry. He is a sensitive naturalist, with good powers of observation, but they lacked direction. He wasn't going anywhere. There is, of course, no reason why he should have been. The aim may have been to capture something of the spirit of a gannet, but where the work is ostensibly biological, direction becomes crucial. Two things helped me beyond measure. One was the assistance of my wife – more of it, I believe, than even Fraser Darling was blessed with – and the other was the opportunity to live for three long seasons cheek by jowl with my gannets.

Routine work, which would have been long and difficult, if possible at all on my own, and which would have been quite out of the question without living on the rock, became easy. It also, incidentally, mapped out my future, but of that, more later. Here, I merely meant to remark that routine work is a marvellous antidote to boredom on a rock or a desert island. Like gathering wild fruit for jam or wine, it gives a sense of achievement without detracting from the enjoyment. This may seem too trite for words, but subsequent experience convinced us that the way to enjoy a totally uninhabited island is to have plenty of worthwhile work to do. Lazing around is bliss for a day or two. Then what? Real life on a tropical desert island is not what you might think. And on a rough old rock in the bleak North Sea, work is a hundred times more important still.

Because it could stifle work, I think weather was the key to our Bass moods, particularly in summer, when it could be unbelievably bad. In mid-July 1962, it was grim. At the height of summer the Rock was damp and dripping, shrouded in dense mist. Limestone never looks as dreary as wet basalt – like a blackened and rusty old hull butting into the sullen North Sea chop. The Garrison walls, apart from the lintels of pale sandstone, are built from the rock of the Bass itself and any covenanter brought to this forbidding place on a day like this must have despaired of his life when he thought of the winter to come. Three weeks later, a near-gale from the east, with torrential rain, flooded the chapel for the second time; in September of the year before the water had reached the level of our floorboards even though we were on stone pillars. The chapel was a lake. Water poured over the doorstep and away down the hill. And no month is free from gales. Take the merry month of May, when all the seabirds have eggs and the rock is heading towards the absolute peak of the breeding year. On the mainland it is that incomparable fortnight or three weeks when the best of late spring and early summer fuse in a riot of bird song and blossom. Slap in the middle, 16 May, came a severe westerly gale that blew the goose-heavy gannets bodily against their inhospitable neighbours. Every bird leaned forward, head down, into the wind. All off-duty partners were (sensibly) away and the orderly parallel rows of incubating birds revealed the incredibly regular spacing of the nests. One or two birds literally braced themselves by

38

digging their half-opened bills into the ground like tripod-legs. Three successive Mays were cold, grey and windy.

Strong wind is like sand at a picnic – it gets into everything. Despite the weather we used to eat outside as often as we could and one day the wind literally blew the butter off our bread. Between 18–25 June 1962 a bitter west to north-west gale or near-gale (what the weathermen call a cold airstream) blew continuously, blasting even the hardy thistles outside the chapel. Flaming June.

Then would come a day that made up for it. I liked best of all the soft light in the calm, early mornings of late spring. On the tranquil sea, patterned with guillemots and puffins, every sound carried clearly. The guillemots rested, preened and bathed, rising on plump, white undercarriages and beating their slender wings. Normally you would see it like a silent film. Today, the wings whirred thunderously. Now and again one flipped down into the translucent depths, wrapped in a pale green envelope of trapped air. The shags were gathering nest material. The most desirable item was the inch-thick hemp rope, coiled massively alongside the lifting crane. With remarkable persistence they queued up to remove it, bracing their spare bodies and leaning back on their great black webs. The dead stems of tree-mallow were responsible for that misplaced persistence by occasionally giving way to long tugging. A difficult but successful struggle is a great reinforcer of behaviour. Why should not the rope do the same? Tree mallow stems are as big as telegraph poles to a shag. They did not merely destroy flight-trim; they often capsized the shag. If it could, it carried them to its nest on foot, but treading on the mallow made it as difficult as lifting oneself in a bucket. The puffins buzzed out from the battlements on their ridiculous wings and the rock-pipits parachuted, singing, to earth, every metallic note dropping as distinctly as a penny on a plate. Breakfast on such a morning was simple contentment.

Up in the gannetry, individuals that I knew by number, if not by name, were fighting, mating, displaying, preening, sleeping and incubating. We became highly involved in case-histories. Colour-rings made all the difference, because we knew for certain that we really were watching the same bird as yesterday, last week, last year or (now) twenty years ago. Take old Hornbill. He is number 5 in colony 6 and got his

39

The hide opposite the observation colony on the NW slope, as it was in 1961 (*above*).

The NW group as it had developed by 1984 (*below*)

name because some mishap tore a piece of his beak-covering (the horny outer layer) loose, on the upper mandible. It curled up like a horn, then eventually fell off and left a deep groove which, fifteen years later, is still visible. We look for him every year and so far he has always returned to his site. I imagine him at sea, in winter, snugly riding out a gale whilst the atrocious North Sea does its worst. It is unlikely to worry him. Then there was Jezebel. She was female 543, in colony 5, owner of a very fine site on a knob of rock that made landing and take-off easy. She left her mate and moved in next door, to a site which seemed inferior to me, but then I'm not a gannet. I suspect it seemed it to her, too, because she often hopped across when it was her turn to go off-duty, greeted her old male briefly (he, alas, was still single) and then used his nice site as a take-off point. But what made her case more unusual was that she not only defeated the female of the male for whom she first deserted her old mate, but, two years afterwards, swapped back and defeated the female who, in the meantime, had taken her place. Fights such as those were the worst of all, for both females were highly motivated. Indeed, one battle was not enough to settle the matter, and a filthy, pock-marked, bleeding gannet would depart, only to return in two or three days, and renew the struggle. In such cases, colour rings are beyond price.

Despite this, ringing was the job we most disliked, not because of the hard work, or the filth, or the bites and scratches, but for the tension due to the even-present chance of causing injury, or loss. To catch adults I used a twelve-foot pole with a running noose, which I slipped over the head of a sitting bird after a slow, slow approach on my belly. The approach naturally alarmed the outermost birds, who became long-necked and anxious, but with care, and at full stretch so as to keep as far away as possible, one could settle the noose softly over its head. Wind wrecked the whole thing. Nothing was more frustrating than to coddle a bird along, pretending not to be interested in it, letting it settle down, making the approach and then – puff – the wind deflecting the noose just when, red-faced and with desperately aching arms, you were about to slip it over. Then the noose would close and have to be pulled back for re-opening. Later I designed a better piece of equipment, with which I could control the noose from base, as it were. But catching the bird was only the beginning. One then had a noosed gannet settling back, uncomfortable but determined, onto its egg or chick. Maybe it was a new chick, in which case it would be lying snugly *on top of* its parent's webs. It doesn't need much imagination to predict the outcome if one simply tugged the resisting gannet off the nest. The chick would be sent flying, either over the edge, or down between nests where it might be stabbed by neighbours, who detest wandering chicks, or at best cause havoc in retrieval. Occasionally one would have liked to release the adult and try another, but couldn't once the noose had tightened. It was a matter of teasing the gannet gently off the nest so that the chick fell safely in the cup. But what about the neighbours? They might have sat uneasily through the approach and the noosing, but could hardly be expected to sit phlegmatically whilst the noosed bird threshed around with a deafening 'urrah urrah' of alarm that could be heard at Tantallon. So it became progressively more difficult and more stressful for all concerned. The actual handling of the adult was not a problem so long as there were two of us. The only tricky bit was getting hold of the gannet which was dancing at the end of the pole, threatening with a beak that can grip and lacerate more than any other bird I know. It is saw-edged, to hold fish and has phenomenal gripping power. To hold a large and muscular mackerel is no easy task, and what can hold a mackerel can hold a finger – or anything else, such as a nipple or a nose or even more personal parts if given half a chance. Once secured, it is just a matter of weighing, measuring, examining for moult, ringing and, if it is a really demanding day, taking a blood sample for biochemical analysis. By the time you've caught and processed a dozen gannets, you've had enough, and the gannets have had too much. I'll come to the justification bit later.

Ringing chicks is hardly less stressful and I invariably emerged filthy, bad-tempered, bleeding and with a splitting tension-headache. I know most gannet ringers do not suffer in this way. Indeed, some even enjoy it, but they do not go about it as we did and their victims certainly don't enjoy it.

Between the nests lies a glutinous black ooze of mud, decayed seaweed, ordure and spilt fish. It may be more than ankle deep and the stench which arises when the crust has been punctured owes its peculiar pungency to the mixture of

42

hydrogen sulphide, fishy amines and ammonia; decaying fish marinated in liquid sulphuretted hydrogen. Agitated chicks regurgitate piles of steaming mackerel or boluses of glistening sand-eels which have to be cleared by hand – my hand – from the nest. We avoided gloves because they hindered sensitive handling of chicks and ringing itself. We squelched stealthily from nest to nest amidst this malodorous slime, in a crouching or even sitting position not from some thwarted childhood desire but out of consideration for the gannets. Often, chicks were best handled without removing them from the nest which thus entailed squatting or kneeling between the nest drums. Nests are about thirty inches apart, centre to centre, and if our cautious progress paid due dividends we had the pleasure of conducting operations within a close circle of highly dangerous adult beaks. Naturally, gannets are extremely agitated by the human in their midst and they show it by a deafening din. The head feathers of males, in particular, stand on end, greatly enlarging their mask. Some bold individuals jab frenziedly and to good effect at any part they can reach. We learnt to assess the mood of our birds very accurately, as well we might. We could judge whether a bird would back off whilst we dealt with the chick, whether it would stand its ground but do nothing or whether it would attack. The out-and-out attackers, almost always males, simply had to be removed. To put our faces near to such males would have been asking for serious trouble. We've had our unprotected face within striking distance of adult gannets hundreds of times, but always of birds that we knew would not attack. But to remove aggressive males is easier said than done. They resist dislodgment with every sinew. And when they are forced off the drum, an unpremeditated departure leaves a trail of disturbance and may even knock smaller chicks off their nest. So it's a last resort and we learnt how to avoid it. For instance, a gannet to seaward is less scared than one whose escape appears to be blocked. Or we engaged the adult with one foot whilst snatching the chick adroitly from the nest, returning it the same way. Sometimes, whilst edging along a ledge face inwards to a vertical cliff, an adult becomes trapped between legs and cliff-face. If he is loath to pass through the portals and you are equally reluctant to face castration there can be an impasse. One leg, lifted backwards and outwards, may then let the gannet out. Hans Kruuk, of later fame among the hyaenas of the Serengeti and Ngoro-Ngoro, once trapped a gannet between his stomach and the ledge – a nasty moment on the east cliffs.

When dealing with large chicks, however, we kept between them and the deep blue sea, so they could be restrained if necessary. It was no good suddenly realising that a large chick, as ill-balanced as a doughnut on a matchstick, was about to leap delicately onto the ledge below, whilst we were stuck on the one above. This meant a bit of route-planning. Things never go perfectly and there are always hideously wilful chicks that scramble away, 'yipping' wildly and spreading panic on all sides. These 'yippers' are a trial, so sore that even to write about them is painful. They have to be caught and returned to their rightful nest, again and again if necessary, and it is simply impossible to do that without creating further havoc. A bit of applied gannet psychology helps here. Simply placed on his nest and left, he immediately throws himself off again. We had to hold him there for a while and then, judging the moment to a nicety, withdraw gently when he had relaxed. If done properly, we had the satisfaction of seeing him preening unconcernedly whilst we worked away only two nests distant. If we left him stranded, thinking he would make his own way back, we exposed him to a grave risk. But the trouble they caused is beyond belief.

Then there were the tiny chicks and unhatched eggs. They could not be left exposed to the weather and the attentions of gulls, but had to be covered with loose nest material. All this was slow and tiring work, but it was the only justifiable way. I have seen ringing parties on the Bass that made my hair stand on end. Striding around in a gannetry, with adults fleeing, chicks scrambling away and herring gulls in full cry after vomited fish and exposed eggs betrays colossal ignorance.

But to compensate for all this, I have often squatted in the midst of a snowfield of gannets, on a fine July evening, immersed in the life and vitality of the teeming colony – gannets flying in, re-united pairs displaying ecstatically almost within touching distance, noise, excitement, smell, beauty. At such times the field naturalist feels highly privileged, perhaps (for a moment) unique. And the feeling owes something to the understanding of all that is going on. It may be hard to see why this should make any difference, but it does. Whereas in music, art or ballet, understanding must surely take you nearer to the feelings of the composer or artist, I could never really approach the feelings of a gannet. But some things do rub off. For instance, when I was in the midst of a vast gannetry the social excitement of the colony was driven home with a sledge hammer and *demanded* attention. Why all the noise and fuss and display? What did it do for gannets? I felt I had to try to answer this sort of question.

5

The Work

Most people do seem to demand a reason, or at least
an excuse, from the man who lives on a small island
without human neighbours, for he is apt to be deemed
asocial and is failing to be regimented.

F. Fraser Darling, *Island Years*

What did we actually *do* for three years on the Bass? The task
of living there was incidental to the main job. This is not the
place for gannet biology – that I have already written – but
spring, summer and autumn were solid with gannet work
and the next chapters are about this. •

When I first set foot there in July 1960 I knew simply that
gannets had not been sufficiently studied. I still marvel at my
good fortune in finding, not only that our most spectacular
seabird was so little known, but a place like the Bass where
we could live and work. And yet many people said precisely
the opposite. It had been the subject of a scholarly and
well-received monograph – and monographs were rare – and
its world population was better-known than that of any other
seabird. But monographs can be long and learned about
nomenclature and history and thin indeed on the vital statis-
tics of ecology and behaviour: James Fisher's long monograph on the fulmar leaves the reader uninformed about
most things necessary for an understanding of their biology
– not through any neglect on his part but simply because the
information had not been gathered. How long do they live?
When do they first breed? How often do they breed? How
successfully? Do they change their nest sites and mates?
What is their social behaviour like? On and on the questions
go, and the answers are not to be obtained except by long-
term work. It was much the same for the gannet. Gurney's

46

account fascinated me but it didn't tell me what I wanted to know, especially about behaviour – my first love.

Julian Huxley had long been keen on gannets and had made an early film about them. Later, he sent me his hand-written notes and it was an encouragement to me to discover that a scientist and naturalist of his calibre wrote disconnected little jottings just like anybody else. Not even the best biological brain looks at a complicated piece of behaviour and quickly understands, records, arranges, interprets and explains. And he had his pioneer study of the great-crested grebe under his belt. Richard Perry's observations, as I have already said, were too vague and unstructured to be of any help. All this was an advantage because it led me to believe that I had the chance to discover things, almost from scratch. In particular, it was clear that 'pure' behaviour had been almost totally neglected, as it still is, in seabird studies. Excluding the Oxford groups, it was almost true to say that, at that time, nowhere in the world had anybody properly documented the *behaviour* of any seabird. Even within the Oxford groups, much of the work was ecologically slanted. I was particularly stimulated by the work of the British Ornithologists' Union Centenary Expedition to Ascension, for it included Douglas Dorward's observations on the behaviour of the masked and brown boobies, relatives of the gannet, and the most superficial glance at the gannet's behaviour demonstrated that the comparison was going to be fascinating. I will have another go at this theme in the final chapter.

The first thing that struck me about gannets was their incredible fights. We set up a hide overlooking a group on the north-west slope. On the cliff-face the gannets nested on tiny ledges, some of them so small that one wondered how the bulky chick managed to keep its place, but from the cliff-top they spread onto the slope leading towards the summit of the rock (before I die I would like to see them nesting on the very summit). The inland edge of this cliff-top spread was merely a muddy strip of no-man's land, beyond which again the hillside was given over to the predatory herring gulls. Just beneath the summit a small inland cliff, actually part of the caldera of the old volcano that spawned the Bass, held about twenty pairs of kittiwakes and two of fulmars. That muddy strip was going to teach me a great deal. It was there that the young males first established the territories for

The NW cliffs from the 'pulpit'. Gannets use small ledges and spread onto slopes or flat ground when the cliffs are full.

which there seemed to be so much violent competition. It is quite wrong ever to use words like 'vicious', 'spiteful', 'cruel' and so on when describing animal behaviour, although some apparently anthropomorphic terms are pretty harmless. 'Affection', for instance, may even be a fair approximation to the feelings between mates – we cannot really know – but 'viciousness' is a singularly human condition denoting a particular kind of awareness and intent which assuredly other animals do not possess. Equally, to describe appalling human behaviour as 'animal' is an ignorant slur on animals. So gannet fights should not be labelled in these terms but they certainly are astonishingly aggressive. More often than not we found that males were contesting sites even when there was plenty of unoccupied ground nearby and this really puzzled us, although I think that I now have part of the answer.

A lot of fights arose between females due to triangular associations, the bane of gannet as well as humankind, and how these birds survived their punishment I hardly know. Such triangles can easily arise. The lone male gannet on his site 'advertises' for a female. This means merely that the male responds to the presence of a female nearby by performing a specific display which is recognised by the female as an invitation to join him on the site. Female A duly approaches; it is for this that she has been prospecting, both from the air and on foot around the fringes of the colony. She may remain on the site for one or two days during which the male mates with her and brings nest material. The pair repeatedly enact the lovely greeting ceremony and preen each other delicately and protractedly – the 'kiss' preen. All this creates a strong bond between them, just as their equivalents do with us. Suppose, having bonded with the male, she now leaves the colony. The male stays behind, guarding his site where, at this stage, he spends most of his time, whilst his new mate oars her way steadily south to the Farnes, across to St Andrew's Bay, up to Aberdeen or wherever her fishing grounds are. Left alone, the male may 'advertise' to a new female. If she responds, and pairs with him, the stage is set for a titanic struggle when the two females meet. There is, incidentally, a good reason why the male may bond with two females and it has nothing to do with polygamy. It is simply that the first few times that a male attracts a female, she is quite likely to

'Kiss-preening' helps to maintain the pair-bond.

wander off again before the bond has been firmly cemented. The male can't *know* that a female is going to return after a long fishing trip until she does so. This is a problem that affects the pair only when the bond is first forming. Once the female has been away and returned a time or two, the male no longer advertises seriously to prospecting females. In the very earliest stages of pair-formation, I saw one male attract no fewer than five females within about two hours. This is clearly one reason why the gannet pair takes a long time to settle down and become a properly functioning unit, and until it does so it has little chance of breeding successfully. It also raises the interesting question of why some females visit an advertising male, but then move on, whilst others proceed to bond with him. There are several possible answers but I do not know which is the right one. A few casual females may be already mated, and merely on their way out of the colony. Again, others may be vacillating because they have not reached quite the right internal state. This could have an extremely potent effect on their behaviour. But it is possible that females evaluate the male and his site. If they do, I have not the slightest idea what they look for. Female choice

Severe territorial fights between male gannets often occur
even before any nest has been built.

certainly exists in the first instance, in that not every advertising male succeeds in attracting her, but whether that is because of anything *he* does, or fails to do, or because of *her* 'state of mind', is impossible to answer.

The reason why two females which are both strongly bonded to the same male fight so bitterly is the same as that which makes two rightful male owners of the same site struggle to exhaustion. It is the extra 'conviction' imparted by ownership. Right is often might in conflicts of this nature. To each of these two females, the other is the intruder. An ordinary intruder soon flees, but a rightful occupant resists to the bitter end.

To return to the males' territorial battles, in that first memorable spring of 1961 it seemed remarkable that such titanic struggles should occur. What was the prize, save a tiny patch of muddy ground, to our eyes indistinguishable from others to be had for the asking? Yet, from our windy little hide we watched one conflict after another. Indeed, most of the colour-ringed males in the fringe had to go through this ordeal. The only suggestions I could find were disappointingly general and unsupported by any evidence, such as that in cases of territorial fighting the struggles may be over the best sites, rather than any old site. Several things could contribute towards a good site – access, drainage, safety from predators among them. Also, and importantly, since Darling's work it seemed that herring gulls nesting on the edges of a colony were less successful than more central birds. It was easy to jump to the conclusion that edge birds were more vulnerable to predation. At that time I did not even know whether edge birds *were* less successful. Later, though, I found that this was indeed so, but it was because edge birds were, by and large, young and inexperienced, and such pairs were less successful whether at the edge or in more central positions. I discovered, also, that fringe-nesting gannets were not more vulnerable to predation, since there were no adequate predators. Man can be ignored, because even if he is counted as a natural predator who could have an effect in evolutionary terms, he would not have taken just edge birds. Furthermore, the competition for sites, which I was so interested in, was precisely *between* fringe males, rather than between males competing for a central site. But in 1961 I had no evidence or answers. I knew merely that males were going to extraordinary lengths in competing for apparently ordinary sites in the fringe of the colony. So the question still remained, why the fighting?

I concluded that even the most admirable site was no good if it was too far from the established group of nesting birds. As I have just said, the natural predators, the herring gulls, were incapable of tackling undisturbed nesting gannets, so they were not the reason. What advantage *was* there in being near to other nesting gannets? Darling had shown that big groups of herring gulls bred more synchronously than small, and were more successful because these groups lost fewer young, in proportion to their size. This was because predation occurred at a more or less steady rate, so that if it had to be endured for longer, more chicks were killed. His observation argued for a social effect of numbers, such that being part of a big group made females lay more nearly at the same time than they would otherwise have done. This social effect of nesting, the 'Fraser Darling' effect, was, I came to believe, vitally important in the gannet, not just for the synchrony which it helped to produce, but because social stimulation made gannets lay earlier than they would otherwise. It works *via* the link between the input from the senses – sound, sight, touch – and the gonads. The frantic noise and bustle of a seabird colony – and anybody who has been in the heart of a great gannetry will admit that it makes Harrods' New Year Sale look like Little Pudlington's Church Bazaar – stimulates each of its members and accelerates the maturation of gonads. By acting with greater force on those who come into the group with relatively undeveloped gonads, it brings forward the date at which the females lay and helps to synchronise the group. Could there be enough advantage in gaining a site which was open to such social stimulation, to explain the fighting? Even if there were plenty of good physical sites there might not be plenty of prime social ones.

I became interested enough in this possibility to try and prove that gannets which were part of a large group would lay earlier and more closely together (in time) than those breeding in a small group. My idea was simply that the former would be exposed to more noise and visual stimulation than the latter. I easily selected two groups of twenty well-established nests in the centre of my main study colony and painted a circle around each group. Luckily, my other

53

study group contained just over twenty well-established nests, in addition to a small number of new sites in the fringe, and it was comparatively isolated. If it had been completely out of sight and earshot of other gannets it would have suited my purpose even better, but I had to make do with what there was. And if this semi-isolated group did lay later, and take longer, then I could assume that a fully isolated one would have shown these traits even more markedly. Ideally, I should have had several groups of each kind and matched them precisely for age and status, but this is asking too much of field work. But the main thing was that my group of twenty semi-isolated pairs were largely old and experienced birds; if they had been largely young breeders the results would not have told me anything about social stimulation because any difference between them and the other two groups could have been due to age.

In both of the two groups from the middle of the mass, the first egg was earlier and the period over which eggs were laid was shorter, than in the semi-isolated group. This result, although in line with what one would expect, fell far short of answering my question, which was why gannets fight so hard to gain a site near to other nesting birds. I still had to find out if earlier and more synchronised laying improved breeding success. And even now, twenty-five years later, I have not proved this. But what direct evidence there is points strongly that way, as also does circumstantial evidence. In most years, early eggs do not succeed significantly better than later ones (provided they are not *very* late) in producing fledglings. But they almost certainly produce fledglings which have a better chance of survival. In other words, it is not in the colony that early chicks are at an advantage, but afterwards. The reason for this is simple. Juvenile gannets migrate to North African waters, or further, immediately after fledging, and they go on their own, for their parents remain behind, still firmly attached to each other and their site. If they run into gales en route, they are greatly at risk – newly on the wing (if at all), running out of fuel (their stored fat), with no experience of feeding themselves, faced with a tremendously demanding hunting technique (plunge-diving) and, on top of everything, on a stormy sea in which even experienced adults could hardly fish successfully. And the chances of running into a gale begin to increase sharply from September on-

54

wards. Since early eggs give early fledglings, it could be *this* advantage that social stimulation provides, and *this* for which such a socially adequate site is worth striving.

Gannets are large and conspicuous and dead ones are often washed ashore. If they are ringed at a known age it is easy to calculate whether they fledged early, late or in between. But unfortunately, ringers usually do not record the age of the chicks they mark. Also, many young gannets die from unnatural causes; they may be shot, trapped in a net, or oiled, and these fatalities should bear no relation to the fledging date. Only deaths related to starvation are relevant. So it is taking a long time to gather enough recoveries of ringed youngsters, of known fledging date, that died of starvation. But the results bear out the prediction, that early fledglings survive best. Sites that produce them are therefore worth fighting for.

This, in essence, is a possible explanation for those ferocious gannet fights. It may not be the whole story – indeed I should be surprised if it were. And it raises almost as many questions as it answers. If I could spend another thirty years working intensively on gannets, I might answer some of them; the easy work has been done and now the questions become more difficult. The whole subject of social stimulation – its nature, costs and benefits – remain greatly underexplored, perhaps simply because it is such a long and difficult task to gather the necessary numbers, weights and measurements, and more difficult to handle.

The easier facets of breeding are of course grist to the mill of the field ecologist. One of them is the extent of the period over which eggs are laid. Before the first one appears, human incursions into the colony are not damaging to the birds although they may well inhibit laying a little. But as the eggs begin to fall thick and fast it becomes difficult to keep up with them. Every egg has to be marked so that new ones can be detected. This means disturbing birds that have already laid and risking some loss of eggs to the ever-thieving herring gulls. Time and again one encounters these insoluble problems – you want to know this and that, and yet simply cannot find out without incurring unwelcome costs. Another practical difficulty, never considered in the enthusiastic planning stage, when it seems possible to do everything, is the use of time; when to do what. This can become a major problem when failure to carry out pre-planned checks are going to leave awkward gaps. My eyes have always been greedier than my stomach and in consequence fieldwork can become a juggling act. April 13th, 1961, was such a day. The morning was spent in the hide watching behaviour; in the afternoon we toiled away building a level boulder base for the hide, on a steep slope. Then came some filthy hours in the quagmire of the western slopes, marking new eggs. Finally, some pressing paperwork, and it was past midnight before I saw my camp bed. Many days were similarly filled and the pile of notebooks grew. Far too quickly when I thought about all that had to be done with their contents. A fieldworker's records are his currency – they are all he has to show for long hours in the field, and the basis of all he hopes to achieve.

Everything is interesting and relevant when trying to get 'inside' an animal. This, at any rate, was the way I looked at it and I became interested in the gannet 'club'. Martin Martin, writing in 1753 about his visit to St Kilda, had described these birds as

> a tribe of barren solan geese which have no nests, and sit upon the bare rock; these are not the young fowls of a year old, whose dark colour would soon distinguish them, but old ones, in all things like the rest; these have a province, as it were, allotted to them, and are in a separate state, having a rock two hundred paces distant from all other; neither do they meddle with, or approach to those hatching, or any other fowl; they sympathise and fish together.

I must admit at the outset that the 'club' defeated me. I know little more about it now than I did after the first few months.

I first became aware of it in the summer of 1960. I topped the rock in a stiff north-westerly and a snowstorm of gannets suddenly filled the sky. Gradually, their ranks thinned as they drifted seawards, and, as far as the eye could reach, the blue surface became dotted with white specks. There must have been more than two thousand. After that, I became accustomed to this great gathering on whichever side of the rock faced the wind. With a mainly northerly blow they gathered beneath the north-facing foghorn, from which I would watch and photograph them. The most obvious features of the 'club' are the preponderance of immature birds in their ranks, the dense but irregular spacing of the individuals, quite different from nesting birds, the constant com-

56

The gannet 'club', a gathering of (mainly) immature birds in which social behaviour is developed.

ing and going, the obviously tentative behaviour of 'club' birds and their greater wariness. They are liable to explode in a panic of madly threshing wings if anything disturbs them, and the panic is irrevocably contagious. Numbers build up gradually from April to July, which is the peak month for attendance, and fall off from mid-August. Also, the proportion of immature birds, especially first-year individuals but to some extent second-year birds too, is greatest around mid-summer. Of the many questions the most critical, it seemed to me, concerned the adult-plumaged 'club' members. Why were they not breeding, or were they merely off-duty birds resting awhile in the 'club'? My attitude to this question shows how easy it is to be prejudiced. Very quickly, I leaned away from the possibility that these adult plumaged birds were in fact off-duty breeders. It fitted my general thesis about the importance of the nest-site and the pair-bond that

57

breeding birds, when present in the colony, should spend their time on their site, and not elsewhere. This is not to say that I ignored contrary evidence but from the outset I had something of an axe to grind. However, the circumstantial evidence did appear to support my position. For one thing, the amount of time spent by the breeding birds on their sites was so great that, as continuous two-day watches of marked birds demonstrated, it hardly left time for anything other than fishing trips. For another, the behaviour of the adult 'club' birds seemed totally out of keeping with that of breeders. They were excessively wary; they were only transiently territorial, 'defending' the position they happened to be in faint-heartedly, easily vanquished; there was a lot of to-ing and fro-ing between the sexes, with low-intensity greeting behaviour. Sometimes there was actual copulation. It seemed unlikely that a bird would be bold on its nest but very wary a few score yards away (as it may have been). But behaviour *is* much affected by context and such a discrepancy is not totally impossible. It is improbable, though, that a male would be highly territorial on his proper site and yet would establish a transient 'club' site on which he would be only faintly territorial. Nor is he likely to perform incipient sexual (pair-bonding) behaviour with one female in the middle, as it were, of a full relationship with his permanent mate. In addition to all that, it is perfectly obvious to the most casual observer that 'off-duty' birds, that is, partners who are not actively incubating or tending a chick do in fact spend a lot of their spare time on the nest, with their mate. So why not spend *all* of it there?

Nevertheless, the identity of these adult plumaged 'club' birds seemed so important that I had to think of some way of establishing it beyond doubt. If they were off-duty breeders there was no special problem, whereas if they were non-breeders there were all sorts of fascinating possibilities. Maybe they were 'unfit' to breed. If so, in what way? Maybe they were experienced breeders taking a rest-year? Maybe they were old 'senile' birds? Did such individuals exist? As I said earlier, I am little nearer to answering these questions. I could not even catch sufficient numbers to see if they turned up among the breeding birds. I had hoped to dye large numbers, say pink, and then look out for them in the colony. One idea that didn't work was to lay a length of perforated hose in the places where 'club' birds congregated and then attach the end of a stirrup pump and pump dye through the perforations. Another idea, no sooner mooted than discarded, was rocket-netting. Too much risk and too difficult in a high wind. Another was dazzling with bright lights at night. I can't remember what happened to that one, except that many 'club' birds did not roost on the rock and anyway the idea of stumbling around on the steep faces or cliff-edges in the pitch dark seemed unattractive.

So I had to be content with merely watching them. At first I was surprised to see that they displayed, skirmished, formed pairs and even mated. I had no good reason to be surprised, but I suppose I had been conditioned, by other people's use of terms like 'loafing areas', to think of 'clubs' as congregations of idle birds. In fact, activity may be one of the main functions of 'clubs'. Even instinctive displays, those which do not have to be learned by imitation but which develop as a result of the birds' genetic endowment, are not necessarily perfect at first. The very performance of them can lead to polishing, both of the physical movements themselves and, presumably, of the perception and processing of the cues to which the behaviour is the adaptive response. In a 'club', one can see immature birds, easily identified by their plumage, responding in a hasty and 'jittery' way to the advances of other individuals. I saw how perfunctory and incomplete several of their displays were, when compared with the same display performed by old and experienced breeders and it seemed plausible that the experience gained in 'clubs' helped to perfect each individual's social behaviour, so vital to life in a gannetry. Of course, it is almost impossible to demonstrate this sort of thing – impossible to prove that a bird which did *not* spend time in a 'club' would not, at a later stage, telescope its polishing process so much that it would be at no disadvantage. But it is at least likely that a gannet which has established its own patch in a 'club', defended it, even though the challenges were never full-blooded, displayed on it, 'advertised' for a mate, greeted her, preened her and so on, over and over again in the two or three years preceding maturity, is better able to do these things in the colony proper than if it had never practised them.

Maybe 'clubs' have no function. Perhaps 'club' gannets simply gather on the Rock for no particular reason and, once there, are stimulated to interact with each other though to no useful purpose. But in that case why should they spend time on a dangerous and 'expensive' rock instead of remaining at sea? Dangerous the Rock certainly is, since the chances of injury are greatly increased by repeatedly landing and taking off in violent and capricious winds. They would be far safer at sea. It is expensive too, in terms of the energy used in flying backwards and forwards and the time and energy spent on land instead of feeding and resting at sea. Darwinism insists that advantages there must be, however slight, if natural selection has produced this shape rather than that, this behaviour rather than the other. And natural selection has produced gannets that spend lots of time in 'clubs' when there is another option. Gannets could easily remain at sea until of breeding age, so far as the ability to survive there is concerned. They can sleep on the sea, feed in it and drink it. So 'clubs' must have a function.

Arguments like this do not usually spring to mind fully fledged as one sits watching gannets. It takes time, observation and the stimulus that comes from applying other people's ideas to one's own work. When it comes down to it, very few people have even a single truly original idea. So much of what appears to be our own thinking is in fact derivative and this becomes increasingly so as knowledge builds up. If I had not read Tinbergen, Lack, Lorenz, Darling and a hundred lesser lights, how much could I have understood about gannets? I wonder. Good, hard data, however, really *are* one's own, and, properly presented, they can rarely fail to be useful to somebody. Still, good ideas are harder to come by than good data.

Eventually, working on my individually recognisable birds, I found that the adult-plumaged gannets in the 'club' were not taking a year off from breeding. All my colour-ringed birds bred, or tried to breed, every year provided that their mates remained with them. If they lost their partner through death, or (rarely) divorce, then the following year they might fail to lay but even so they remained in the colony. So that was one possibility accounted for. This was not surprising for as soon as I could show that gannets retained their sites and mates from year to year it became unlikely that they would vacate the site and abandon the mate for the joys of the 'club'. If rest-years *had* been essential

Adult gannets were ringed for individual
identification (*above*); June fixing a ring (*below*).

it would have been a much better ploy to 'rest' on the site, especially since its acquisition had cost them dear. This 'jig-saw puzzle' effect – each bit of information dovetailing with the others – is one of the soundest reasons for a broad-based approach, especially if it lasts for many years.

When Darling worked on herring gulls hardly anybody fully realised the enormous potential of long-term studies using marked individuals. Only Richdale, the dour and defensive schoolmaster in a New Zealand backwater was really getting down to it. Year after year on Otago Head he ringed, weighed, measured, recorded laying dates, fledging dates and success rates, etc., of his yellow-eyed penguins. It seemed obsessive but it triggered off a spate of fruitful studies, just as Darling's work did. In fact Darling's herring gull has become one of the most-studied birds in the world. There has been a positive Niagara of ecological and behavioural investigations, some of them of a sophistication that would have greatly impressed him. Questions asked and answered have included: does the position of the egg within the clutch (laid first, second or third) influence its success in producing a fledgling; what relationships are there between the spacing of nests in the colony and their success; which kinds of spacing are most attractive to new recruits and where do these come from; what effect does altering the density and size of the colony have on the age at which new recruits enter and breed, and many more.

But, more and more, such studies require teamwork or full-time professional work, often with considerable paid help. It seems that the day of people like Fraser Darling is drawing to a close. Mass-participation enquiries are also favoured, and very productive they are, too. But I can best identify with Darling's approach, bivouacking in the glens and camping on his lonely islands. He achieved a great deal, unaided.

The 'club' birds were much warier than the breeders. Early in the year even the newly-returned breeders were hair-trigger scary compared with their boldness later in the season, when they would often stand their ground and even attack a human intruder. But in February or even March, as soon as a human figure came within sight, however distant, they became alert, long-necked and anxious. We used to crawl along the hillside on our bellies to get to the hide. Often

62

a single bird even warier than the rest precipitated a whole-sale panic simply by its own agitated departure which, later in the season, would have had no effect whatsoever. I am sure that it is such individuals that were mistakenly thought to be the sentinels by the St Kildans:

> The solan geese have always some of their number keeping centry in the night, and if they are surprised, as it often happens, all the flock are taken one after another. But if the Centinel be awake at the approach of the creeping fowlers, and hear a noise, it cries softly, Grog, Grog, at which the flock move not; but if the centinel sees or hears the fowlers approaching, he cries quickly, Bir, Bir, which would seem to import Danger, since immediately after, the whole tribe take wing.

But most revealing of all, so far as the birds' internal state was concerned, were the sudden, inexplicable 'dreads'. Periodically and completely without outside stimulus, the whole group fell uncannily quiet, as though at a signal, the harsh hubbub of guttural calls fading and dying. For a suspense-filled moment the colony held its breath. Then, with a roar of wings the whole mass surged pell-mell towards the cliff-edge and the safety of the sea. Gradually, these panics became less frequent until, some time before the eggs were present, the gannets were never troubled by them, the noisy commerce of the colony continuing unabated from before dawn until after dusk. But by mid-August the first signs of re-awakening wariness were detectable. The control is internal, the wariness a manifestation of the hormonal state. Significantly, the attendance of 'club' birds corresponds with the period during which breeders are most firmly attached to the colony. By late August many 'club' birds, especially the younger ones, have gone.

It is worth returning to this tremendously important matter of internal state later, but writing of tameness reminds me that it is itself often difficult to understand. The Bass gannets that frequently saw people approach, pass by and disappear without causing them to leave their nests, became astonishingly bold. Those that were regularly approached *too* closely soon began to panic long *before* they were closely approached. The critical distance had been greatly increased. This is simple conditioning. One would therefore expect that birds which are rarely or never closely approached would be tame.

63

Yet on Ailsa Craig, where gannets are highly inaccessible, they are much warier than on the Bass. Easy, you say. The Ailsa birds simply are not conditioned to accept people, so they panic. But what about the Galapagos, where many of the seabirds live on totally uninhabited islands and yet are notoriously tame? I have walked up to frigates and boobies and lifted them from the nest and replaced them. Once I placed a red-footed booby on top of my wife's head and it simply sat there. No doubt it can all be understood in terms of the precise past-history, long and short term, of the individuals, and long-term of the populations. In this respect, the experience of Sarah Wanless, at the time a research student of mine living on Ailsa Craig, is fascinating. In order to climb and follow the course of egg-laying in certain sheer cliff areas she put permanent safety ropes on the face. In the areas thus visited, birds returned later and laid slightly later in the two following years, than those round about them, even though climbing visits had been discontinued.

There are constant reminders of the enormous role played by this invisible but potent internal state. Take any behaviour – fighting, territorial display, mating, the greeting ceremony, 'comfort' movements such as shaking the head and wings – whatever it is it shows its own seasonal pattern, which is due to internal state. I spent untold hours simply counting all these behaviour patterns in standard groups, and in birds of different status, all through the season, so that I knew how frequently they were performed. It is surprising what can emerge even from such a simple procedure. For instance, it showed that the territorial display, a dramatic and highly stereotyped set of movements very easy to see and to count, was frequent early in the season, scarcer in mid-season and very common again later on. The lovely greeting ceremony followed exactly the same pattern, as also did the ritualised 'threat-gape' by which neighbours menaced each other, thus maintaining the necessary space around their nest. This suggested to me, and still does, that these displays shared a common internal factor which 'caused' them. That they were all 'instinctive' behaviour patterns, in the sense that they developed in their precisely predictable form as a result, not of learning in the sense of copying others but by virtue of normal development encoded in the genes, could hardly be doubted. The ethological literature is full of such

64

Ritualised threat maintains the distance between neighbours.

innate behaviours and controlled experiments have shown that they do appear even when the possibility of copying has been stringently excluded. So it seems that a group of quite distinct, innate displays are all partly controlled by the same set of internal factors. This may seem a trivial conclusion, but it is relevant to the controversy which has raged, sometimes acrimoniously, about whether the concept of innateness in behaviour ought even to be used and, in particular, about the possibility that a particular set of behaviours which serve the same major function (in the gannet example used above this would be the territorial or spacing-out function) can be under the control of the same 'drive'. Some distinguished ethologists become apoplectic at the very idea.

Another small but interesting finding was that the males' territorial display was more intense than the females'. It was the same display, but a more vigorous version. Whatever internal factors, no doubt largely hormonal, helped to cause the display were acting more potently in the male. Whilst this was what one would intuitively expect, it was nice to see it happening in such a measurable way. And it led to the prediction, again fulfilled, that young males, new to the colony, and those in 'clubs', would also perform a low-intensity version of the territorial display, again partly due to (presumably) a lower level of the male hormone which we know to be implicated in aggression.

All this tempts me to add my widow's mite on the subject of human behaviour. I will resist, except to venture that it seems to me immensely inconsistent to accept that man is a mammal, that male and female mammals always show differences in behaviour as they do in physique and physiology, that in other mammals these behavioural differences are partly inborn and most certainly are not solely learned and yet to assert that in man, alone among primates and mammals, there are no inborn differences in behaviour between the sexes, only 'learned' ones. If there are innate differences, ought we not to know about them, understand them and if necessary and desirable, as it may well be, modify their effects by appropriate social and cultural means? Yet, to hear some feminists talk, not to mention male fellow travellers from the wilder ideological and sociological shores, it is contemptible even to investigate the matter, much less to accept findings that may be in some sense unpalatable. One encounters the same antipathetic reaction towards any suggestion that intelligence is partly inborn, or that the different human races differ in aptitudes and abilities, as though the very possibility somehow implies all sorts of nasty things. It is precisely to guard against exploitation, political, ideological, or whatever, that we need to know if, how and why there are such differences. I would hazard a guess that the overwhelming majority of biologists who have worked long in the field, with whole, live animals, would reject out of hand the notion that human sexes and races differ in behaviour only because society has conditioned them to do so.

However, to return to gannets, another display that prompted many hours of watching and counting was a bizarre posture-cum-movement that earlier observers had found both amusing and fascinating, and to which several of them had not hesitated to apply an injudicious measure of wishful thinking. This display was at once comical and dignified, as displays so often are. Many ordinary movements, when grossly exaggerated, become comical. The effect of a gannet stretching its neck and pointing its bill to the heavens, whilst bringing its piercing eyes to focus, binocularly, forwards and downwards, at the same time swivelling its wings to raise the tips, depressing the tail and parading with high-lifted, droopy webs, displaying their pale green lines, lacks only the accompaniment of some suitably ludicrous sound to complete the impact. This is provided when, eventually, the gannet launches itself from the cliff, for then it utters a sepulchral groan, or a disyllabic, higher-pitched 'oo-ah', still in this looped-up posture. It is no wonder that earlier observers noticed it and, in fair agreement (though with an element of copying) concluded that, since it was performed prior to departure from the colony, and departure, if the gannet had to travel through the nesting ranks in order to reach the edge, was often resented by the disturbed neighbours, the display was appeasement behaviour. The rather nice term, 'excuse-me', was coined for it. The gannet, by removing its bill from any possible attacking position, was said to be signalling its harmless intention of passing through its neighbours' space. To complete the story, it was necessary that the behaviour be effective. It would be hard to describe this bizarre display as appeasement behaviour if in fact it failed to appease. This is where observation failed and imagination triumphed, for more than one distinguished ornithologist said, in black and white, that neighbours *did* allow skypointing individuals an unmolested passage to the fringe.

My gannets did not behave at all like that. Time and again I saw birds posture for minutes on end but then, being still threatened by neighbours, they abandoned their display posture, put their heads down and ran to the edge, sometimes pecked and hassled on all sides as they went. It required more than the eye of faith to see skypointing as effective appeasement. Indeed, and quite understandably, it actually provoked attack even before the posturing bird moved off its site. The neighbours inevitably recognised skypointing as the prelude to intrusion and reacted accordingly. But, equally clearly, the function of skypointing could hardly be to elicit aggression from the neighbours. Yet it certainly conveyed the message 'I am about to move off my site'. To whom was the message directed and why? This is what I meant earlier by suggesting that birdwatching is both more enjoyable and more interesting when it has an aim.

I collected getting on for a thousand samples of skypointing, recording the circumstances of each occasion. It soon became evident that the intended recipient of the message

A pair of Australasian gannets, with two-week old chick, greeting (*above*) and changing-over at the nest (*below*). The left-hand bird is 'skypointing' to indicate impending departure. The black secondary and tail feathers are not found in the Atlantic gannet.

was the mate. Perhaps the most instructive occasions were those upon which *both* partners skypointed simultaneously. This didn't often happen, but when it did, it revealed the function of the behaviour. Both partners may skypoint 'at' or 'to' each other, but only one of them actually departed and it was always the one which had postured longest and most intensively. The mutual posturing may have been maintained for more than a minute, but always one bird gradually backed down whilst the other continued at full stretch and departed only when its mate had quite relinquished the posture and clearly, therefore, no longer intended to go. And this was the whole point of the display – to make quite certain that only one partner left the nest and that the other stayed behind. The departing bird had to ensure that its mate accepted its intention and was prepared to stay behind. The reason was obvious. An unguarded nest invited neighbours to pilfer nest material and they were quick to accept. As a result, an egg or small chick would be in grave and immediate danger of being thrown out, or, equally disastrously, taken by a predatory gull. In less than a minute after the nest was left unguarded, it could have lost its contents. This was reason enough to evolve a signal to avoid that danger which, potentially, could arise whenever the partners changed-over to allow the on-duty bird to depart to feed and its mate to take over the care of the egg or chick. The gannets were using a ritualised display to achieve a rapport which they have neither the intelligence nor the vocal capacity to achieve by 'talking', but which was vital to them. This is the main function of visual display in animals, whether of honeybees communicating the whereabouts of food or a male stickleback enticing the female into the nest.

I may appear to have made a song and dance about nothing. To the reader, the problem may not have been apparent. People may assume that a gannet knows it mustn't leave its egg unguarded. They may invest birds with rudimentary reasoning powers. But this is not how behaviour at this level works. By the same reasoning, a gannet would be expected to help its displaced chick back onto the nest, or at least to feed it off the nest, but it does neither. It either attacks the chick, or lets it starve. After reunion with its mate, the appropriate behaviour for the relieved bird is to go to sea to feed, and the only requirement is a timing mechanism to ensure that the departure occurs at the right time. The skypointing display is such a mechanism.

Displays raise another fundamental point, which is that concrete proof of their function is unobtainable. How could one prove beyond doubt that skypointing synchronises nest-relief and that omission of the display would lead to some loss of eggs or young? By comparing change-overs that employ skypointing with those that do not. But this is impossible because skypointing is standard behaviour during nest relief. This does not invalidate the conclusion. After all, I cannot compare the success of gannets that incubate their egg under-foot with those that hatch them in another way, but I do not doubt that the function of incubation underfoot is to hatch eggs.

The amazing efficiency with which animals cope with the complexities of their lives, yet without rational thought and discussion, leaves only two options. Either natural selection is the most wonderful shaper of adaptive behaviour, or something else is. The religious person prefers 'something else', but I never understood why. Should it be thought in some way more desirable that an unimaginable entity – God, or 'Ultimate Reality' – suddenly 'created' a beautiful and effi-cient bird, which we call a swallow – than that evolution produced it, over millions of years? It is as though even the glimmerings of an understanding of how things came to be what they now are, renders that understanding distasteful. To be beautiful life has to be mysterious. But of course it is so complex that it always will be mysterious. We can never empty the rainbow.

Julian Huxley elevated his understanding of evolution, in its most advanced, man-centred form (his 'psycho-social' evolution) to a visionary concept. Man would become the directing-force of his own evolution, he would 'select out' undesirable traits, not by crude eugenics but by applying an understanding of the process of culture change. But culture includes the spiritual dimension. Huxley consistently advo-cated the pursuit of self-awareness by which we possess a potential for feeling and seeing in ways that transcend mere responses to knowable stimuli. He didn't feel the need to reject all that he could not understand, but he certainly never rejected understanding. Indeed, in his attempt to accommo-date the 'extra dimension' that he believed to be a desirable

70

if not essential part of the whole person, he sometimes went beyond that which more orthodox scientists would allow. For instance he accepted much of Teilhard de Chardin's *The Phenomenon of Man*, which Peter Medawar described as vague and incomprehensible. It certainly was to me.

The point of this digression was merely to recognise that evolution produces the most fantastic adaptations and that these are behavioural, just as much as physical. Animals cope with the incredible complexities of their lives, in the absence of thinking as we understand it, by means of unbelievably acute reception and integration of clues from the environment (which includes other animals), in conjunction with a fluctuating inner state about which we understand rather little in terms of how it meshes with external stimuli. Skypointing is one of a battery of signals by means of which gannets are able to regulate their actions, in this case with reference to their partner. I will describe other parts of their sign language later.

6

Sex and Violence

All aspects of modern evolutionary biology [which includes behaviour] can be seen as part of a research programme inaugurated by *The Origin of Species*. It is without doubt the most important biological work ever written.

    Richard Leakey

As the bird books describe it, the gannet's life-history seems simple. But this, naturally, is deceptive. Of course it is easy to say that it returns to the rock in January, lays in April, takes 43 days to hatch its egg and 91 to rear its chick, and so on. But the fine texture of its life is indescribably complex.

Take copulation, for instance. An enduring Bass memory is the sight and sound of gannets mating in the raw, misty days of late winter. The male, mounted on the female's back with his wings spread like an heraldic eagle, seizes her nape with a pinch like bolt shears. The skin and feathers are all scrunched up between the tips of his mandibles and the small dislodged feathers float softly down to the mud. His great webs, toes clenched for grip, thump loudly on her back with muffled, hollow smacks that carry far on a still day. When he dismounts he leaves a large muddy patch – a useful telltale. He takes anything up to half a minute, which is a phenomenally long time for a single copulation. Herring gulls may take longer but the male remains on the female's back for several successive inseminations, or at least cloacal applications. But there is much, much more to say about gannet copulation. It is in itself an essay in social behaviour, as I will try to show.

Males do not initiate mating; that is the female's prerogative. She solicits simply by shaking her head vigorously from side to side, whilst sitting. This headshake is interesting; it

72

A pair of Australasian gannets mating. The male grips the female's head and tramples with his webs. The preen gland is exposed at the base of the tail.

crops up in several gannet displays. Outside displays it is the normal reaction to a wide range of touch stimuli such as water or dirt on the head or bill, and since most displays involve touching the ground or nest material or another gannet, they can readily evoke headshaking. But the *route* by which headshaking has become the female gannet's soliciting behaviour is not explained as simply as that; a devious detour through scenic behaviour country is in order.

The soliciting headshake grades over from headshaking used in greeting – that is, greeting headshakes gradually become soliciting headshaking in some circumstances, especially when both partners are sexually excited. This applies particularly to the period before the egg is laid and even more to the time when the pair-bond is being forged. Perhaps the best place to begin tracking down our quarry, the soliciting headshake, is with a male gannet newly established on his site. He is standing (not sitting) in the centre of his little patch, at roughly jabbing distance from his nearest neighbours. A fight develops nearby and the combatants, as often happens, encroach on several neighbouring territories in the explosive violence of their encounter. Our male, among others, is agitated by this intrusion and he begins to perform his 'site-ownership display' which tells other gannets that he is a site-owning male and warns them to keep away. This particular display is one of the commonest sights in a gannetry and the source of much of the strident hubbub, for it is contagious and accompanied by harsh calling. At first sight it may seem to be merely a bizzare series of movements but remember that it is sign-language, watch it a few times and it becomes obvious that there are two perfectly regular main components. The most spectacular of these is a sweeping downwards movement of the head, so extreme that the ground may be touched or even bitten. This is repeated several times interspersed with vigorous headshakes. The wings are held out, angled at the 'wrist' and the display terminates with a conspicuous arching of the neck which presses the bill tip hard against the upper breast. These components form a complete unit and all of them are tied together in the same order – and no matter how often the

The head-flinging component of the 'bowing' display (*above*) followed by the actual bow (*centre*), after which the display finishes with a bill-tucking (appeasement) posture (*below*).

74

display is repeated (sometimes the colony is in an uproar for hours) the whole package is performed each time. This site-ownership display could occupy years of observation and still reveal new insights. But the component which concerns us at the moment is the headshake and how it became part of copulation. That it has become part of site-ownership is not difficult to understand when one sees that the downward sweep sometimes involves ground-biting. The ground is muddy and so the bill becomes soiled, and headshaking dispels the dirt. The ground biting, incidentally, is symbolic biting, agression directed at a substitute object. So far so good. Headshaking as part of the aggressive site-ownership display seems understandable.

Now things quieten down and our male is standing in a relaxed posture, head retracted, on his site. Examined closely through binoculars his eye movements reveal that he is evidently taking a keen interest in all that is happening around him. There is a constant traffic of flying gannets a few of which land at the edge of the colony. Some of these are young unmated females (how often have I wished that there was some way of recognising the precise status of flying birds!). The male gannet, though, can at least recognise them as females, which is more than the human observer can do.

This close resemblance of the sexes is an important point. Birds (and also fish), rather more than mammals, react strongly to certain conspicuous features of their own species, not bothering about details but latching onto the key stimuli. Even a lousy model of an owl will bring out the mobbing response in a chaffinch. We may take it that an approaching female gannet is emitting strong 'male' signals, simply by looking like a male. In other words, if our male gannet's eyes and brain were totally like a computer they would tell him that something very, very like a male was about to intrude. The adaptive reaction to intrusion, as I have just described, is the site-ownership display (bowing). But watch the male closely. He does not bow. He headshakes in a highly exaggerated manner with slight forward and downward reaching movements. He is, in fact soliciting her, 'asking' her to approach but – and it is an important but – *he is using the site-ownership display* which she has necessarily triggered by looking like a male. To be more accurate, he is using *part* of the site-ownership display. The *aggressive* bit (the downward

76

sweep to bite the ground) has been reduced to that slight forward reach (an 'intention' movement) and the headshake, which is merely a modified head-cleansing movement without aggressive connotation, has been greatly exaggerated. This seems to me very clever! He is allowed, by this modification, to be both true to his own 'feelings' (motivation), which are to declare his ownership of the site in the face of potential intrusion but, vitally important, he is able at the same time to convey a quite *different* message to the female. He needs to say 'come here', not 'go away', yet the part of him that would say 'go away' is, by this device, allowed expression. This is very important because it keeps his response to potential intrusion nice and strong.

Now we are nearly back to headshaking in copulation, which is where all this began. As she approaches in response to his soliciting display he headshakes more vigorously and she, too, begins to headshake. They meet; he attacks her; she appeases him by turning her beak away and they begin the lovely greeting behaviour. Now she sinks into a squatting position headshaking so vigorously that it is almost a flinging movement – the extreme form of soliciting headshake. She raises her tail, he grips her nape, perhaps letting go momentarily to bite her vigorously and mounts. His webs begin to tramp faster and faster until at the instant of ejaculation he bends his tail under the female's and applies his cloaca to hers. After straining briefly, immobile, he relaxes, steps off her back and 'skypoints', signalling that he is about to leave the site. Usually, he flies off to gather nest material.

From all this, the headshake which the female uses to solicit copulation is seen to be strongly involved in sexual behaviour but also in aggressive behaviour, whilst its original context was merely that of cleaning the head or bill, a use to which it is still put. It has been used as a building block in complex displays during which it has come to signal emotions or 'tendencies' which are *not* present in the simple cleansing context. Of course, things may not always happen exactly as I have described; much depends on imponderables such as the precise motivational state of the male. For instance, he may invite the female to approach but when she does get close he may switch to vigorous attack, her male-like appearance tipping the balance that way. She always tries to weather the assault by facing away but in this she may fail even if she

77

endures the punishment for ten or twenty minutes, a long time to withstand such a battering. Many a time I marvelled that she, who would attack an intruding female with a total commitment not a whit inferior to the male's, nevertheless meekly accepted such harsh treatment just because it was handed out by a male. But the stakes are so high that natural selection has ensured that she can and does. If she were to retaliate there would be no chance of forming a pair-bond and bang would go her prospects of reproduction, which brings me to the role of nape-biting during copulation and to the male's trampling movements and the exceptional duration of the act.

In the gannet, a long-lived and monogamous seabird, the pair-bond is very important and copulation can powerfully reinforce it. If, somehow, the male's aggressive nape-biting can be linked, in the female's feelings, with sexual behaviour, then at one stroke the female's tolerance of her mate's aggression is increased (which benefits both her and him) *and* they acquire a potent means of reinforcing their bond. So the male, by biting the female during copulation, actually enhances its effectiveness as a pair-bonder. If copulation is long-drawn-out this effect is even greater. If he increases the tactile stimuli even more by tramping up and down on her back, better still.

So copulation is not merely a means of fertilising the egg but is an integral part of the gannet's pair-behaviour. I do not think that this is a 'just so' story but is an interpretation which takes full account of the facts and is evolutionarily plausible.

Copulation begins early in the year, long before the egg is laid. By the time it is beginning to enlarge, the female has a strong bond with her mate and usually a good nest. She then goes off to build up her reserves. This exodus of females is not dramatic but it becomes obvious that most of the nests are being guarded by lone males. He seems not to have the opportunity to recuperate from the arduous weeks already spent guarding and displaying on his site. If the female goes, he must stay, since the nest would be pilfered within minutes if left unguarded. This is very different from the fulmar, which deserts the breeding colony *en masse* before egg-laying.

The snell winds of winter still fingered the Bass when the first gannet eggs appeared sometime in late March. The

78

freshly laid egg was easily identified because its rough limey covering still bore fresh traces of blood from the oviduct and had not been stained and dirtied by the gannets' muddy webs. Thousands used to be collected from the Bass and sold locally as 'one of the most rare delicacies of the season', highly appreciated at the Royal Table (that was before Prince Philip's time) and (quite optimistically) 'admitted to be indistinguishable from Plover's eggs'.

All the early eggs were in large, well-built nests; none were in the sparsely furnished mud-patches of the outer fringes, for these belonged to younger birds and they invariably laid late. In fact, it takes three or four years before a given female reaches the laying date which she will thereafter stick to, more or less, for the remainder of her life providing she doesn't change her mate. If she does lose him or change him, she tends to lay later for a year or two afterwards. On rare occasions we happened to be in the hide (that's one advantage of thousands of hide-hours) when one of these young females laid her egg and the first few minutes were both interesting and revealing. Two or three of them simply didn't know what to do with it. They stood up, prodded it, picked it up and dropped it and eventually either lost or broke it. Several others probably did the same for their eggs disappeared in the first day. We never saw a 'first-time' male do this. Some of these inexperienced mothers managed to incubate their egg but lost it during hatching, and again the manner in which they did so was very instructive. Normally, the parent which is on duty when the egg begins to hatch (usually, I believe, the female) transfers it to the top of the webs for the obvious reason that if she didn't her weight would crush it. But a few new mothers failed to do this. Instead they continued to incubate it underfoot and killed the chick, leathery though it is.

I find this particularly interesting for three reasons. First, it suggests (as I try to show in my account of the young gannet making its first flight) that there are finely timed developments going on within the gannet's nervous system and until these have reached precisely the right point the appropriate behaviour simply will not occur. Second, some individuals reach this stage earlier than others and third, there is no question of intelligence operating to ensure the right response. I need hardly say that I have no idea what

79

The gannet incubates its egg underfoot and removes hard
lumps from the floor of the nest to avoid puncturing the shell.

these internal processes are, but neither has anybody else.
Nevertheless, they are real and although, in humans, they
can be substantially overridden, they occur there, too.

Another surprising thing, at least superficially, is the lack
of flexibility in the behaviour of most birds and certainly of
gannets. It was deeply frustrating to watch a gannet with its
egg or small chick slightly displaced. The simplest thing in
the world would seem to be for the adult to pick it up gently
in that most precise and skilful implement, the bill, and put
it where it should be. If the out-of-place object is a scrap of
nest material no effort is too great. The gannet will try again
and again, with unfathomable patience, to pick it up with the
tips of the mandibles, so delicately that if it were a goldcrest's
egg it wouldn't break. But its own egg or chick does not
register as 'nest material' (why should it?) and does not
trigger picking-up behaviour. The gannet does not, and
presumably cannot, 'think' about it, and use its skills in a
novel context. Certain behavioural acts are inexorably tied to
certain situations and cannot be transferred. Sometimes particular acts are tied to ludicrously precise situations. Often,

it isn't even a question of using behaviour differently. For instance a displaced chick may be only just outside the nest but apparently the gannet not only cannot pick it up and put it back (which would be 'new' behaviour) but cannot, or does not, even *feed* it in this slightly off-stage position. Feeding would not involve any new act but simply the normal one directed a few inches to one side. So, if the chick is too small to crawl back, it dies. This lack of flexibility, which looks so like stupidity, is absolutely typical of the gannet and is a consequence of relying on genetically determined behaviour to deal with potentially complex situations which can often pose new problems. Instinct works extremely well, but rigidly. If it were to incorporate flexibility it would become a different ball game with different rules and might be called intelligence. Gannets are not intelligent. Nevertheless, as intelligent animals ourselves, we cannot help feeling astonished. I have seen young Abbott's boobies make a bad landing in the jungle trees where they nest and place themselves in dire danger of falling to certain death. Yet they have not, and could not have, used their bills to *grasp* a twig or creeper and pull themselves, parrot-like, to safety. All they could do was to use the bill, far less effectively, as a lever. Grasping was simply not a function which the bill could perform in that situation. Yet that same bill is used to *grasp* and tear living twigs off the tree for nest material! Gannet and booby behaviour is perfectly adapted to a vast but markedly finite set of circumstances. The rules governing the development of the appropriate nerve-muscle pathways *and the central directives which activate them* remain a major problem of biology which, for my money, will never be fully solved.

To return to the Bass, by the time the chicks were half grown it was high summer and the Rock was bursting at the seams. There were now three gannets to every nest, to say nothing of the thousands of kittiwakes and the guillemot chicks. Between the gannet nests every hollow was littered with flight feathers and the small body feathers drifted and swirled in the breeze. To my surprise, the adults began to moult as soon as the eggs hatched. One might have thought they had sufficient demands on their energy feeding that insatiable youngster, without the additional burden of replacing feathers, but the mackerel were by then inshore and it was probably easier for the gannets to moult at that time,

The chick (five weeks old) has partly exposed its web;
this helps it to lose heat.

regardless of their chicks' demands, than at any other.

These were the days to relax and soak up the sun, if there
had been time. Fraser Darling was a wise old bird and knew
that hours spent lying idly on the cliff-top, watching the
many activities of the seabirds, was not purely a luxury.
Half-formed thoughts could firm up and new ideas be born.
Whenever the sun shone hotly I used to steal an hour or two
in a sheltered nook. Sometimes I took some work but it was
never properly done. Sunbathing is a serious pursuit and

doesn't mix easily with labour. The young gannets had no choice in the matter, and suffered severely in the heat. In the lee of the rock, exposed to a baking July sun, soaking up the heat radiated from the black basalt and clothed in blubber and thick fluff, their only defences were to flutter the bare throat skin and to stretch out their webs to radiate heat, doubly effective if they first excreted on them to enhance evaporative cooling. It took me some time to wake up to that trick. At first I couldn't see why they so often stuck out a web in warm weather, nor why their feet were so dirty. Often they conserved energy by laying their heads on the ground like sleeping dogs, which made them look quite dead. I often heard visitors 'poor thinging'.

It was in these days of early July that we reluctantly donned our ringing garments and resigned ourselves to a few hours of filthy and exhausting work. Each year we put colour rings on many of the chicks so that, when the few survivors returned, we would know how old they were. We didn't have long to wait. Within less than two years, occasionally even before the first year was up, we began to spot them. Of course they were still wearing immature plumage. The second-year birds were boldly pied above and the first-year-olds still largely brown, although already white beneath and 'capped', with a white collar breaking up the brown on the neck. They turned up in the 'clubs' that I described earlier, among the adult-plumaged 'club' birds and those with only a few dark feathers remaining and two intriguing facts quickly emerged. First, there was clearly a strong tendency for immature 'club' birds to choose temporary partners of their own age. These pairs soon broke up but even so this peer-attraction seemed significant. This particular observation could just as easily have been made without benefit of colour-ringed birds but the next one couldn't. Age for age, females retained more dark feathers than males, so that in colour ringed pairs in which we knew that the partners were born in the same year, the female *looked* younger.

In another year or two colour-ringed youngsters started turning up in our observation colonies. This was even better, for it promised us breeders of known age – an invaluable asset. Fraser Darling could have done with some, but he never stayed with his work quite long enough to make it worthwhile. There is no substitute for knowing individual

The lighthouse colony at its beginnings in 1961.

identities beyond doubt. It soon became plain that young breeders were attracted to their own bit of the Bass Rock but were not wholly bound to it. There was, for instance, a spur at the base of the east cliffs which, with thirty or forty nests, was full up. Chicks from this spur turned up as young breeders in the expanding areas on the cliff top above their birthplace. One such area was directly opposite the top of the lighthouse tower. When we started work there were less than ten occupied sites in that group and we colour-ringed the chicks each year. Now there are more than 500 sites. If you were to calculate the output from those original pairs and assume that the adults remained faithful to their group throughout their lives (which is true) and that *all* the surviving young returned there to breed it would show that this increase could not conceivably have taken place without an influx of young breeders from outside the group. The same thing happened even more dramatically in our main observation colony, which ran to around 100 occupied sites in 1961 and has now snowballed to a vast concourse of more than 3,000.

There was food for thought here. First, why did my two

The lighthouse colony as it had developed by 1984.

study groups increase so disproportionately? The answer seemed fairly straightforward in that they happened to be on parts of the Rock that were not only suitable for colonisation but also were protected from disturbance. To some parts of the Rock neither of these things applied and so they did not attract recruits to the same extent. But it was evident, all the same, that prospecting birds responded to the presence of young, site-owning birds by settling and joining them and that the process snowballed. This immediately dovetailed with a feature of gannet behaviour that had slightly puzzled me. For hour after hour, hundreds or thousands of gannets sailed around the Bass, but not, as it seemed, aimlessly. They were closely inspecting the colony. Round and round they went, some individuals stalling and scrutinising a particular area time after time. Now, I believe that they were doing one, or both, of two things. First, some of them were almost certainly inspecting the very spot on which they grew up and fledged. If there is a real opportunity of acquiring a site near to that spot, upon which it is imprinted, I believe a young male will, more often than by chance, take it. Second, they were assessing the social characteristics of particular areas. I mean that they were registering, for instance, the status of the birds below – were they old-established pairs or was the area largely occupied by young pre-breeding birds? There could be several advantages to joining the latter group. For instance, a prospecting female would be more likely to find a receptive male among a gathering of new site-owners than among old-established pairs. Conversely, such a new male would have a better chance of gaining a mate if he were in an area to which young females were attracted. Again, a young pair among old-established ones would be far behind them in laying and it is much better to be more or less in step with neighbours. This is because interference from neighbours is reduced when you and they are at more or less the same stage of the breeding cycle. There is a lot to be learned about this and it is one of the fruits of Darling's work that the advantages of synchronised breeding are now claiming attention.

So I was fortunate to place my hide near to a rapidly expanding group, for there I could watch territorial and pair-forming behaviour proceeding at full pitch. Sometimes we could hardly wait to get back into the hide to see whether a keenly contested site had changed hands again.

But the very success of my groups in sucking in recruits from other parts of the rock posed another problem. If gannets would go to a new part of their natal rock would they not go to a completely different colony? This was a matter of major interest although there was extremely little direct evidence. But there was one good source of indirect evidence. If a colony grew faster than its own output could have ensured then, obviously, it was gaining recruits from elsewhere. Many years later I calculated that some colonies such as Grassholm had done precisely that, although the Bass Rock's increase, at just less than three per cent per year, was about what would be expected from its own production of young. Of course the Bass is the only colony on the eastern side of Britain, apart from the small and relatively recent colony on the mainland cliffs at Bempton, whereas there is a whole string of gannetries up the western side. Since some of these western gannetries swap young recruits it may very well be that there is a large population of pre-breeders plying the sea-lanes between them, perhaps even spending time at more than one gannetry before finally settling for good. The Bass on the other hand is, I suspect, largely autonomous, apart from donating a few recruits to Bempton. But all this was far from my mind when I first colour-ringed my young birds, for then I merely wanted to know whether they would return to their own little group and at what age they would breed.

By the time I had begun to think about the wider problems of interchange between colonies I was given the means to look at it, in the form of a scholarship from the Natural Environment Research Council (NERC) to finance a student on Ailsa Craig. In this way Sarah, in her turn, was fortunate enough to spend three seasons on this marvellous Craig in the Clyde, 'Elsay . . . quherin is one grate hill, round and roughe, and one heavin [haven] and an aboundance of Soland Geise'. Her home was quarryman Jimmy Girvan's old cottage. I chose Ailsa partly because it was the only gannetry in Britain, or anywhere, that had been counted (by Jack Gibson) every year for more than twenty years, but more significantly because those counts appeared to show substantial, indeed sometimes massive increases and decreases in successive years. To cut a long story short, it seems that although, over the *longer* term, Ailsa's population has simply

87

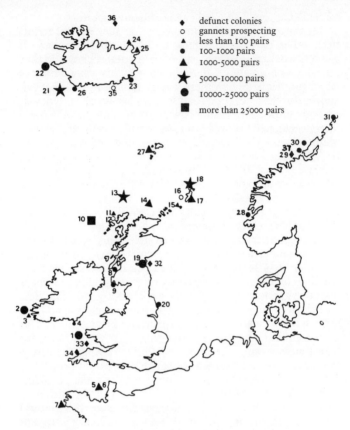

The distribution of the gannet on the eastern side of the
North Atlantic. The Norwegian colonies recently (1985)
totalled approx. 2,300 nesting pairs. A new colony at Engoy-
holmen is forming. *Key:* 1 Grassholm; 2 Little Skellig; 3 Bull
Rock; 4 Great Saltee; 5 Ortac; 6 Les Etacs; 7 Rouzic; 8 Ailsa
Craig; 9 Scar Rocks; 10 St Kilda; 11 Flannans; 12 Bearasay;
13 Sula Sgeir; 14 Sule Stack; 15 Fair Isle; 16 Foula; 17 Noss;
18 Hermaness; 19 Bass Rock; 20 Bempton; 21 Westmann
Isles; 22 Eldey; 23 Skrudur; 24 Raudinupur; 25 Stori-
Karlinn; 26 Mafadrang; 27 Myggenaes; 28 Runde;
29 Skittenskarvholmen; 30 Skarvlakken; 31 Syltefjord;
32 Isle of May; 33 Lundy; 34 Gannet Stone; 35 Ingolfshofar;
36 Grimsey; 37 Hovsflesa.

increased at much the same rate as the gannet population in Britain as a whole, it has shown much larger increases in *some* years and also considerable decreases at least in certain areas of the Craig. These substantial decreases between one year and the next are particularly interesting. Increases, whether large or small, pose no problems. Moderate ones can be accounted for by the productivity of the colony and larger ones by an influx of recruits from elsewhere. But, assuming that they are not just counting errors, what on earth can the decreases mean? They certainly had no counterpart on the Bass. Were they non-breeders, not holding 'proper' sites and therefore footloose enough to move on the next year? Their identity remains to be solved. There is some evidence that Grassholm sometimes decreases from year to year even though the overall trend is upwards and it may be that there is considerable traffic between gannetries on the west coast but very little between east and west. But good estimates of *all* the west-coast gannetries in several successive years is a tall order indeed. At the moment the lack of such a series and the potential errors in the estimated numbers (errors due to counting difficulties and to the differences in the numbers attending the colonies at different times of year and of day) make detailed interpretation hazardous.

In fact, though, the general story both here and on the other side of the Atlantic is of a species which has suffered a marked decline and is now steadily increasing. The cause of the decline was obviously the sustained and often murderous pressure exerted by killing adults and young for food (or, in Canada, for bait) and collecting eggs. This largely ceased before the end of last century and the steady recovery began. Nowadays, only Sula Sgeir among British gannetries is culled. This privilege is accorded to Lewis men because of their ancient and continuing tradition.

It is a moot point whether the present upward trend of the gannet owes anything to an increase in food supply; indeed it is not known whether there *has* been an increase. On the contrary, several food-fishes, particularly herring, but also mackerel, sprats and sand-eels, have been heavily fished by man or are in the process of being. But the gannet population continues to thrive, as indeed do several British seabirds including the puffin. Several new gannetries have been established in the last twenty years, including one on Fair Isle and

Grassholm, in the Bristol Channel, is a comparatively low island for a gannetry, but it is well situated near rich fishing grounds; the gannetry has grown explosively this century.

Foula, and the Norwegian colonies date only from the early 'sixties, while a little further back, gannets first nested on Scar Rocks (south-west Scotland) in the late 'thirties, in the Channel Islands in about 1940, in Brittany only a little before that and on Great Saltee, in Ireland, around 1930, about the time that the only British mainland colony, at Bempton, also started. Both these last two teetered along with two or three pairs for many years before they finally took off and began their upward climb. The main Icelandic colony is Eldey, 'the mealsack', a flat-topped volcanic rock which, since its protection from 1940 onwards, has grown to somewhere within the 'teens' of thousands of pairs.

At present there are some 34 to 37 colonies of the North Atlantic gannet in the world, all but six of them on this side. It is difficult to be absolutely precise because the situation is always changing so far as the small colonies and those very few which may be on the point of starting up, are concerned. There are only fifteen colonies holding around or over 5,000 'pairs' (better described as 'occupied sites', most of which are owned by pairs but a few of which belong to single males). In Britain and Ireland the biggest, perhaps approaching 50,000 occupied sites, is the St Kilda complex of Boreray, Stac Lee and Stac an Armin. Then comes Grassholm, Ailsa, Little Skellig and the Bass, not necessarily in that order but all approaching or over 20,000 occupied sites. Incidentally, every gannetry has its own atmosphere and excitement and it

A view of the Grassholm gannetry in 1975, from the summit.

is not always mere size that counts. I yield to nobody in my
admiration for Ailsa but unhesitatingly claim that the breathtaking vista, the magnificent sweep, of the gannet snowfield
on Grassholm or the north-west slope of the Bass, make a
greater impact – perhaps even than the magnificent Boreray,
though that would be a bold claim indeed. There, against the
sheer scale of the precipice, the gannets dwindle into dust-
motes. But stand on the edge of a mighty congregation,
undwarfed by natural features, though drawing as much
atmosphere from sea and cliff as the great herds of wildebeest
draw from the plains of East Africa, absorb the vitality and
excitement of these strident masses, and you will never forget
what a gannetry is.

Bird Rocks, Gulf of St Lawrence, Quebec, Canada, used

to be the world's greatest gannetry. Lying in the Magdalen archipelago, Great Bird, where the main colony still exists, is not an impressive lump, a mere thirty metres high. But the birds! Surrounded by the icy and often rough waters of the Gulf, with immense fish stocks, the Bird Rocks were, in the words of the earliest account 'as full of birds as any meadow is of grass . . . a great and infinite number of those that we call Margaulx, that are white and bigger than any geese' (Hakluyt's story of Cartier's voyage of 1534). I calculated that, if the decrease which occurred between 1860 and 1864 was accurately gauged, 30,000 adults and presumably many young were killed each year. This seems more than one could possibly believe, but that the slaughter was colossal is beyond doubt and this magnificent colony had all but disappeared by 1881. A paltry fifty nests, all recently robbed, were all that remained on the top of the island. For comparison, the latest (1980–81) census of the number of Australasian gannets in the New Zealand colonies is 46,000 pairs in twenty-six colonies. The African gannet has not been properly censussed recently but in 1963 the population was estimated at 353,000 individuals in some six colonies.

If I cannot tell you *exactly* how many gannetries there are I obviously cannot produce a figure for the number of gannets in the world. But for general purposes I suggest that there are at present between 200,000 and 250,000 occupied sites at the world's Atlantic gannetries. Not all of these will hold a pair, so one cannot simply double the figure to give individual birds. But, allowing for that and also for the number of immature birds in existence as I write (January 1985) I don't mind sticking my neck out and hoisting a figure of something around half a million individuals. As an order-of-magnitude guesstimate I don't suppose it is too far out and it puts the gannet way down the league table of seabirds, compared with many auks, penguins, tubenoses, gulls, terns and others. That is not too surprising because Atlantic gannets have, after all, a very restricted distribution.

7

Catching the Jump

I think I could turn and live with animals;
  they are so placid and self-contained;
Not one is dissatisfied – not one is demented
  with the mania of owning things.
  Walt Whitman

From the summit of the Bass, looking over the massed ranks of gannets, it seems hardly possible that not so long ago we used to walk to the very edge of the cliffs without disturbing a single bird from its nest. We had a marvellous, flat out-look ledge, called the Pulpit, below which the cliffs fell a sheer 250 feet. It was a grim precipice, deeply shadowed, with tiny ledges and cracks full of ancient debris; the black basalt slimy and green with algae. The strident voices of incoming gannets were amplified by the cliff-face, which curved slightly like a giant megaphone. I used often to lie there, looking straight down onto the seabird traffic, enjoying their mastery over the up-draughts (though once I saw a gannet blown head over tail). One day I saw a nest containing twin chicks, which was most unusual. In fact I have never known a gannet to lay two eggs in a clutch. Replacement eggs, yes, but a two-egg clutch, never. This is not to say that it never happens; only that when it seems to, there is usually another explanation. Occasionally, two females lay in the same nest and I described earlier how these triangles of one male and two females can arise. I had never really wondered why gannets lay one egg instead of two, even though shags lay two, three or even four and seem to do very well. I cannot honestly say that, left to myself, I would ever have asked, but as so often happens, my mind was jogged. My old supervisor, David Lack, displayed his special ability to ask simple but

93

basic questions and to propose answers. His answer on the question 'What determines clutch size' was simple: a bird lays as many eggs as it (and its mate, if he is involved) can care for, and which produce as many chicks as it can feed. There was an important additional proviso. It is no good rearing a large number of young if most of them die before they have a chance to breed. It is the number that survive to breed that matters.

The testing procedure was simple. I merely took newly laid eggs from some pairs and added them to existing fresh eggs in the nests of others, thus doubling up the clutches. That part may have been simple, but the rest was not. I could have left them alone and recorded how many of the twins fledged successfully, but then I would not have known whether they were mere bags of bones, with little chance of survival, or fine, fat juveniles well fuelled for the long migration to North African waters. And even if they were fat, what about the cost to their parents? Had they worn themselves out labouring excessively to meet the huge demands of two voracious chicks? 'As greedy as a gannet' is not without foundation. It would not be a good bargain, in the long run, to produce two chicks for three or four years and then die, when you could have produced one chick for ten or twenty years.

To answer these questions I set myself three tasks: first, to weigh the twinned chicks frequently and see how they grew compared with normal ones; second, to see how often the parents of twins came with food and whether they spent the normal amount of time resting on the nest and third, to weigh the parents at the end of their chick-rearing to see if they were lighter than they should have been. Weighing chicks sounds all very well on paper but practice and precept are often poles apart. Once the twins were old enough to leave the nest and scramble away, we were in trouble. If only we could have tethered them. If we tried to recover them when they ran, we simply created havoc amongst the neighbours. If we retreated, we left the chick to face possible death when it tried to get back to its nest. The bigger they grew, the greater became the problem. Eventually it became insuperable, which is why our records of twin weights terminated at 70 days instead of continuing until the twins fledged, which is usually at 91 days of age. Even at 70 they are as big as

A fully-grown young gannet ready to fledge. It weighs considerably more than its parent because of the store of fat that fuels its journey to Africa.

adults, considerably heavier and very mobile and our colony was near the edge of the cliff. Our final technique illustrates the high technology of fieldwork in the 'sixties. First, we put up a small tent, divided into two halves, far enough away from the edge of the colony. Then we carefully entered the colony to seawards, so that as we approached the unfortunate sets of twins they would have to stay put or scramble uphill away from danger. We then pounced and whilst June prevented them getting back to the colony I scooped them up, two at a time, and bundled them into one half of the tent. At last, red-faced, sweating and with fluff up our noses, we emerged from the tent with armsful of gugas, dumped them at the edge of the colony and fled. Once we were out of sight they regained their nests comparatively easily because they

95

were moving towards them, slowly (downhill, which is when they go slowest) and not through hostile adults. This was not a task we felt like repeating too often, but we managed it without causing any deaths, our own included. Just how much distress and suffering one imposes on animals is, as things stand, largely a personal matter, but I would unhesitatingly throw overboard a great mass of animal research which causes pain and suffering on a scale quite incommensurate with the knowledge gained.

The second task was easier. It merely involved continuous 48-hour observations from the hide. Here, yet again, the job would have been impossible alone. Finally, weighing the adults meant catching them, and we managed only a few, but enough.

The result of all this was something of a surprise. It seemed that Bass gannets took the extra feeding burden in their stride. The individual twins, up to the time we stopped weighing them, were only slightly lighter than single chicks – sometimes they were heavier – and the parents were usually able to spend substantial periods on the nest together. One of the parents is normally always on guard, but if the effort of finding enough food had been excessive, the other would not have been expected to stay around, too. Nor were they lighter than they should have been.

Once, when I talked about this at a meeting, somebody objected to my conclusion on the grounds that had every nest on the Bass held twins instead of singles, they might all have starved. This entirely missed the point, which was that *some* pairs could apparently manage twins. The question was why did they not attempt to do so? Of course, had they done so, and the proportion of twin rearers within the colony increased, there could conceivably come a point at which those with only a single chick would do better than those with two. Their chick would survive whereas most of the twins, if they were underweight because too many parents were rearing twins, might die. In such an event, the pendulum would swing back and favour single chicks. But apparently *no* gannets were attempting twins even though at the time there seemed to be ample food.

Perhaps the likeliest explanation is that the number of gannets in relation to their food has changed, in the gannets favour, but the one-chick habit, adapted to a leaner time, has

96

not caught up. If we are around, and interested enough, in tens of thousands of years, we may find that some gannets do lay two eggs. On the other hand, if food becomes scarce for gannets due to overfishing by man – by no means an impossibility – they may never get round to it. At the moment they are increasing nicely at around three per cent each year. Whether this is due entirely to the relaxation of human pressure, formerly severe in some areas, or to a change in the numbers and distribution of fish, as a result of changes in sea temperature, or to both, we do not know.

Whilst on the subject of gannets and man, a frightful and largely avoidable danger to them lies in the synthetic line and netting lost or thrown away by fishermen. Gannets, being unthinking creatures, pick it up and carry it to their nests. They even quarrel over it. Sometimes the piece is so large that two or three neighbours can tuck it into the structure of their nests, where it becomes immovably embedded. At some gannetries half to two-thirds of all nests contain this lethal material. The outcome can easily be imagined. The chick, idling away the long days, plays with it. Perhaps it wraps around a leg, or slips over its head. Naturally it hasn't the sense to slip out of it. By fledging time it may be hopelessly anchored, its leg dislocated or cut to the bone. On the Bass, up to seven chicks have been seen roped together by this rot-resistant, and immensely strong material. Adults get caught too, and die wretchedly, hanging from their nest or dangling over a ledge. And gannets are not the only seabird sufferers. Thousands of auks are drowned each year after becoming entangled in netting. Lost netting cannot be helped but if fishermen knew the consequences of throwing it overboard they might get rid of it ashore.

The fledging problems facing the young gannet do not include predators. First, they have to reach the edge of the cliff, if they are from more inland nests. Then, once in the air, they have to cope with their heavy bodies, often in the tricky and turbulent winds that play around the cliffs. But before that, they must cross an invisible threshold – again, that all-important internal state. In 1961 Niko Tinbergen and I were shooting our film on gannets, using a battered old hand-cranked camera from the Oxford Zoology Department. We planned to capture the life of the gannet throughout the breeding season, from first return to final departure,

97

and I was keen to photograph the fledgling leaping off the nest and over the cliff. Not surprisingly this hadn't been done before, since the chances of having the youngster in the camera sights at that precise instant are thousands to one against. Niko always used the camera as a tool, but he undoubtedly enjoyed the hunting and positively gloated over the beauty of the captured instant. He was seldom happier than when editing his work on a little viewer, film cuttings festooned everywhere amongst the tanks of sticklebacks and piles of yellowing reprints. I became as hooked as he was. Were I single, I think I would like to be a wildlife cameraman, like Hugh Miles, Alan Rootes or Dieter Plage. Years later, Anglia Television's 'Survival' series screened my film of the beautiful Abbott's booby of the Christmas Island jungle.

The great satisfaction for the ethologist-cameraman is the documentation of complex behaviour patterns from before the beginning, until after the end. Almost always, the cameraman who is not a naturalist, or not deeply familiar with his subject, begins his sequence slightly *after* the behaviour has begun. The bizarreness, or obvious interest of the behaviour catches his eye and he swings onto it, thinking he's got it. But he hasn't, because he missed the beginning. To get that, you must anticipate. It is simple to get the sort of shots that people go 'ooh' about – magnificent portraits and close-ups showing every feather – even flight shots, so impressive in slow motion, are bread and butter stuff. It is much harder to catch a male gannet just before he begins his inconspicuous little 'advertising' display to a prospecting female. You have to recognise the female and spot the interested male and perhaps pan back and forth between them.

But how to film fledging? On 31 August 1962 I spent all day behind the camera and missed every fledging by a fraction. I knew that several chicks were due to take the plunge. If I concentrated on one, I might choose the wrong one. If I swung back and forth, I risked missing them all. Some were nearer to the camera, and there was no zoom lens. The film was in 100-foot rolls, and if I used it on abortive sequences I risked having no film left in the camera for the climax. Yet if I didn't film the moment *preceding* the jump, I missed what I wanted to capture. And there was precious time flowing past. This dilemma has always made me envy the profes-

98

sional. He has all day and every day to spend on filming. I have always had to squeeze it in with fieldwork, and the two don't mix. Far too often I dragged a mountain of impedimenta with me, through Galapagos scrub and Christmas Island jungle, simply in order to have it there when, my real work finished, I could begin filming. It is as though Hugh Miles had to start weighing, measuring, counting and observing after his day's filming. But I digress.

The young gannet, ready to fledge, has to surmount a barrier none the less formidable for being internal. The void is beneath. For three months he has never moved off his secure pedestal. Often he has braced himself, crouched and clung like a limpet when threatened by gales or by the clumsy landing or departure of an adult. For a cliff-ledge youngster, a slip would have meant death. Now, something inside the young gannet is forcing it inexorably towards the instant when it will reverse all that and voluntarily launch itself into space. This struggle is written large in its behaviour. One moment it may be relaxed, preening, fiddling with nest material, perhaps mildly threatening the juvenile next door. Abruptly it turns towards the sea and begins to peer fixedly downwards, long-necked, apparently oblivious of everything around it. The wings flick rapidly, as the brain commands them to open and fear countermands the order – a struggle by the command systems for the control of the muscles which will execute the behaviour. There may be convulsive swallowing. A juvenile in this state is a tempting target for the camera. Surely it is about to go. I crouch behind the lens, willing the black youngster to jump, perhaps filming several abortive sequences. Then, abruptly, it relaxes, turns its back on the siren call and settles down again. One day I spent thirteen consecutive hours at the lens. My eyelids began to twitch as I peered. Then they literally twitched reflexly whenever I even *thought* of looking through the lens. At this stage, I packed up, and the youngster promptly fledged. But the next day I immortalised number 5002 as it struggled through unsympathetic neighbours and fell like an old sack over the edge of that awful precipice.

The same day I saw a definite bird-of-the-year on the wing. It must have been from one of the more northerly gannetries, for newly fledged gannets cannot raise themselves from the sea once they've landed. They may fly several miles whilst

Take-off (*above*) and flight (*below*). The webs are used to depress the tail and increase lift.

they still have height, but eventually they strike the water and then their fatness prevents them from taking off again, although they often flap enthusiastically over the surface. The adults do not go with their young nor in any way force them to leave the nest. They do not 'starve' them, as so many bird books still aver. It is amazing how tenacious this ground-less assertion has proved. I simply cannot kill it. It seems to appeal to people. The young, having laid down substantial layers of fat, do tend to demand less food during the few days before departure, but that's all. If they beg, they may be fed, right up to the hour they leave.

In August, September and October the North Sea and North Atlantic is dotted with these black juvenile gannets at first swimming and later flying, south. Frequently they at-tach themselves to passing adults and even pester them for food. Perhaps, exceptionally, they may be lucky enough to trigger a feeding response, but their parents, back in the gannetry, are still firmly attached to their nests and beyond question are not feeding their offspring during their south-ward migration. This absence of post-fledging care in the gannet contrasts strongly with gulls, terns and auks and indeed with all the gannet's close relatives, the boobies. Abbott's booby feeds its free-flying juvenile for nine months or even more.

But of course the gannet *couldn't* feed its fledged young at the colony because the fledgling can't fly back to the nest and it would be out of the question to feed it at sea. An adult couldn't be expected to locate its youngster, which would be swimming actively, amidst fog and strong currents, whilst its own foraging trips would inevitably take it scores of miles away. The real question is why have gannets 'chosen' to pro-duce fat-laden fledglings which cannot (at first) fly, rather than leaner, lighter young which could fly around in the breeding area, as gulls and boobies do, and drop back in to be fed. The answer for once seems fairly straightforward and lies in the nature of the gannet's food. As my experiments with twin gannets showed, food is plentiful. Give a gannet two young instead of its usual single chick and it rears sturdy twins. So it is possible for parent gannets to feed mackerel (an uncommonly nutritious fish) to their chicks at such a rate that they grow fast and accumulate plenty of fat. This enables the young to get away from the breeding areas and shoot south in double-quick time, living on their reserves as they go. In this way they are clear of hostile northern waters when autumn gales set in. The gannet's way of doing things ('strategy' is now the word) is simply not an option for the boobies, which do not have either equivalent food or weather. Since a high percentage of juvenile gannets die on this journey, it may seem inefficient, but the main thing is that it works well enough to ensure the survival of gannets. It is the gannet's evolutionary answer to the conditions of the northern seas (to which it came from warmer waters in earlier times) and it passes the acid test – it works.

Some fledglings never get as far as the open sea; they crash. August and September at a gannetry bring a dismal crop of accidents. On 25 August 1961 my diary recorded four fully feathered youngsters at the foot of the sheer north-west face. One seemed in good shape, exercising its wings vigorously. Jammed into a nearby crack was the dishevelled carcase of a fifth bird, fouled with droppings from above. But, horribly, it suddenly reared its head and began to extricate itself. Eventually it heaved clear and dragged itself towards the others, flopping repeatedly onto one wing. Both its legs and one wing were shattered. It was the totally impartial face of nature. The vigorous youngster attacked the cripple, a response which, however repugnant to us, simply epitomised the rigidity of instinct. A young gannet is programmed to repel an intruder which would compete for food and space. The fact that the response was inappropriate here was a minor accident which in no way invalidates the usefulness of the behaviour. Gannets don't think; they act, and nature cannot provide complicated sets of instructions for every conceivable circumstance which might render a generally useful response inappropriate in one particular instance. Even with our capacity to think, we often act inappropriately because of inner promptings which are hard to ignore and which many of us believe are part of our inherited behaviour and not simply the consequences of the way our mother treated us, essential to survival under the circumstances in which ninety-nine per cent of our evolutionary past has been spent. After all, recognisably hominid remains go back at least three million years – some claim twice that – and for how long has modern man existed?

At the base of the west cliffs, an apron of rock protrudes

into the sea. It is a gannet graveyard, a death trap for young-sters leaping down from directly above, a grim place of stagnant, slimy pools and rotting carcasses. The greater-blacks are often there, hunching around on their flesh-pink webs. Fortunately, the cliffs elsewhere fall sheer into the sea, otherwise it might be as bad as Ailsa, where the cliffs in places have rounder shoulders and more extensive, boulder-strewn bases. Here a tremendous number of youngsters, and more surprisingly adults, fall to their deaths, as Gwynne-Vevers described nearly fifty years ago. Later, during her three years on Ailsa, Sarah Wanless had the sickening task of killing hundreds of fallen gannets, having to do her grisly rounds every two or three days. Ailsa swarms with rats and injured gannets are eaten alive. Only a few were in good enough shape to be saved and there is a limit to the number you can feed, so there was only the one answer. Those hundreds of humane killings demanded a great deal of cour-age, for they went on and on. I know I wouldn't have liked to do it.

Gannets are not easily killed. A ringer on the Bass acciden-tally knocked a five-week old chick from the top of the east cliffs. It belly-flopped into the sea from more than two hun-dred feet and burst like a ripe plum. Fred picked it out of the water and handed the miserable, dripping bundle to me at the landing. I borrowed a curved needle and some horse-hair from the keepers' medical kit (these were do-it-yourself days) and sewed it up, thinking to give it at least a slight chance. Ten minutes later it was standing up in its box, thrusting forward its brawny, black-laced chest and flapping vigorously, to the grave danger of its stitches. Later, I gave it to a nest which had lost its chick and it flourished and fledged normally.

By late October or November the gannetry was a sorry sight. Exceptionally late chicks, coming either from replace-ment eggs or from late-laying females breeding for the first time, lingered on forlornly. Almost all the adults had left and on a wet and misty November day the gannetry was a black and soggy morass of disintegrated nests. In the flatter areas the widely scattered youngsters, no longer inhibited by hos-tile neighbours, wandered more freely and many came to-gether, a gathering of the forsaken. They were not quite forgotten, though. Occasionally, at this time, a small skein of adults would oar steadily towards the rock, separating in the usual way when close in, a couple bearing to the east cliffs, three to the west and so on. On 30 October 1961 there were just a dozen youngsters on the rock. The neighbouring chicks with whom they had spent part of the summer were in the Bay of Biscay or off Morocco, long ago, or, just as likely, their decaying corpses were feeding sandhoppers in the tide wrack of some lonely beach, or disintegrating at sea.

Some of these black youngsters survived starvation, fishermen's nets and hooks, guns, oil and more bizarre threats such as swallowing lacerating objects, or being eaten by large fish. Among them were a few of our colour-ringed chicks. We were always on the look-out for them and when they returned there was great excitement. One memorable bird, still immature, brought its rings back to the very ledge on which it had been reared but it couldn't stay because its parents were still there. Usually, as I have mentioned, they established sites on the fringes of the group in which they were reared although sometimes this was not possible. One chick from a group at the base of the east cliffs, where there was no fringe room remaining, settled on top of the east cliffs. A few moved round the rock a few degrees from their birthplace. Two or three even forsook the Bass and turned up in the young gannetry at Bempton, in Yorkshire, the only mainland colony. The Bass birds of course fly past Bempton on their passage North and a few simply become drawn into the colony there.

The very first colour-ringed bird of mine that I spotted was just over a year old, on 10 July 1962. It is rather unusual for a bird to return within its first year and it may be in its fourth or even fifth year by the time it does so. Unfortunately, one can know only when it was first *seen* and never when it first returned. Some go further south than others, and stay there longer. It is extraordinarily difficult to prove why they go south, but the circumstances suggest that they do so because feeding conditions are more suitable for young and inexperienced birds. Presumably the gales, heavy seas, low temperatures and short days of the northern winter would not suit them, whereas adults, experienced, capable of surviving for long periods without food and gorging rapidly when opportunity offers, sometimes on powerful, deep-swimming prey, can cope. By moving south young birds also avoid competing

104

with adults, though this may not be as clear an advantage as it may seem. Gannets choose to feed in flocks; their plumage and behaviour attracts others, who stream in from afar when they spot diving birds. To suggest that younger birds benefit by avoiding these fishing parties is tantamount to saying that it works better for adults than younger birds. This may be so; evidence is extremely difficult to come by.

8

The Other Seabirds of the Bass

It [Bass] is about one English mile [from the shore].
Herein are kept sheep and some kine and coneys;
abundance of fowl breed here, Solem-geese, storts,
scoutes, and twenty several sorts of fowl, which make
such a noise as that you may hear them and nothing
else a mile before you come to them.

Sir William Brereton, 1635

Gannets dominate the Bass, but even without them the old
Crag would be a notable seabird station. The new year begins, not in January but in November or even October, by
which time guillemots may already throng the ledges almost
before the last dejected young gannet has abandoned the
colony. But they are fitful visitors, easily persuaded to leave,
and December and January can be desolate times, when the
Rock broods, eyeless. Things don't really get under way
until February, a bleak month. The herring gulls are there
before then, but they straggle across the grey Forth every
night to roost on the mainland. The first kittiwakes, fresh
from their sea-wanderings and ill at ease, appear in the
middle of the month. They come and go, noticeably subdued
and stay for only brief periods. The passionate crying which
will echo and re-echo in the caverns of the Bass has no place
or function on these February ledges. But by the time the
puffins appear in mid-March (occasionally late February)
the kittiwakes are noisier and the Rock is already alive and
kicking, its sight renewed even if not yet bright-eyed. They
congregate in large rafts offshore, sitting quietly on the sea
for hours at a time. Now and again birds from the ledges fly
out to the rafts as though losing courage on land. Later the
rafts become noisy and excited gatherings, often in turmoil.

106

Cackle and headwave display in the fulmar.

By mid-March the fulmars are well into their oil-spitting territorial battles. They are mysterious birds in many ways, nobody really knows why they suddenly began to increase so spectacularly, racing from island to island and headland to headland earlier this century until they had colonised Britain from St Kilda to beyond the south coast. Only in recent years, thanks to patient work on Eynhallow, in the Orkneys, by Aberdeen University, have we learnt that they may be eleven years old before they begin to breed and that they may live more than forty years. Their behaviour has always puzzled me, for it seems so limited compared with the varied repertoire of displays which gulls or gannets employ. Fulmars use a markedly stereotyped headwaving and cackling display to cope with almost all social interactions. Does a rival male fly up to and past a site-occupying bird? Cackle and headwave. Do the members of a pair meet on the site? Cackle and headwave. Closer scrutiny shows that there are subtle differences in the headwaving in different contexts, but they *are* subtle. It makes do with substantially fewer discrete displays than many other seabirds, yet it has to meet as many communication requirements as the gannet and it obviously succeeds in doing so. Why should gannets or gulls have evolved a wide range of displays whilst fulmars rely on subtle variations of a single theme? For this interesting question I have no ready answer.

The fulmar has not yet colonised the Bass in large numbers. It arrived in 1926 around the old Battlements and still shares those time-worn stones with the puffins and shags, peering from the interstices, across the tree mallow, to Tantallon, where fulmars also breed. By 1961, when we went to live on the Bass, there were around seven pairs in the Battlement walls and about the same number on the eastern side of Cable Gulley, the side that runs up to the headland which looks across to the south-west cliffs. These were the two main areas although there were a few prospecting birds elsewhere and a couple of pairs on the inland cliff, among the kittiwakes. Fulmar numbers are difficult to pin down for there are enormous seasonal changes and sizeable numbers of casual visitors. However, there was a marked increase in 1966. The Battlement group mustered around a dozen pairs and raised at least six chicks whilst the east sidings did even better, with at least seventeen pairs. In addition, birds were patrolling the southern slope where at least one chick was reared. At the time of writing the east sidings population has increased to around thirty pairs rearing some fifteen chicks in 1983, and there are five or six on the inland cliff, the kittiwakes having cleared off. There are several pairs scattered around elsewhere but the Bass population probably numbers less than a hundred breeding pairs, which is not exactly a spectacular increase over nearly sixty years.

The herring gulls are the guardians of the Rock, restless spirits wailing and yodelling all night long. Many's the time I have lain awake, listening to the demonic laughter and the ringing alarm calls. Sometimes all hell broke loose, goodness knows why, for there was nobody up there and no predators except the gulls themselves. Somehow I have never grown fond of herring gulls, fine birds though they are. Rapacious creatures, cannibals, scavengers and habitués of filthy rubbish dumps, sewage outfalls, docks, canals, promenades and parks, places where no 'real' seabird would be found. Quite unfairly, its yellow eye gives it the cold look of a reptile whereas the dark one of the screaming kittiwake imparts a beguiling and quite misleadingly gentle and intelligent appearance. The kittiwake seems a 'clean' gull, working the

108

A herring gull feeding its day-old chick.

waves rather than the refuse tip. All such judgements and comments are of course thoroughly spurious, but still they spring to the mind.

Among the pale grey, flesh-legged herring gulls were three little enclaves of its smart relative the lesser black-back with its deep yellow legs. The first of them arrived before the end of March, despite their predominantly migratory habit.

The appearance of gulls' eggs in the last day or two of April or the first of May marked the welcome onset of fresh eggs and chips. They made fine Yorkshire pudding too. Long before the middle of the month we and the keepers had gathered four or five hundred, but the tide still rolled in and we could not stem it. The notion that because gulls are scavengers their eggs are somehow polluted defeats me. Some people refuse mackerel for the same reason. I don't know what the difference is between an amino acid derived from a drowned kitten and the same one derived from chicken feed.

A huddle of guillemots. There is one bridled bird on the right.

Despite all the collecting, by mid-August the ugly brown juveniles, all beak and sloping forehead, crept everywhere between the tussocks. Caught unawares these large youngsters crouched motionless, although fully able to fly. This behaviour could have been a carry-over from those long weeks during which crouching had been their only defence. Or perhaps they were still slow to take flight and more likely to survive by lying doggo than making a clumsy attempt.

On the Bass guillemot chicks fledge in late June or early July. Nothing captures the atmosphere of seabird cliffs better than the weird ensemble of guttural, gargling guillemots and demented kittiwakes. The pear-shaped guillemot egg, large and beautiful with its bold markings on blue or green shell, which enable individuals to recognise their own egg (an essential ability on a crowded ledge where there are no nests) is laid in April. It was once believed that the guillemot 'solders' its egg to the ledge. William Harvey, writing about his visit to the Bass in 1641, says

> among the many different kinds of birds which seek the Bass
> Island for the sake of laying and incubating their eggs, . . .
> one bird was pointed out to me which lays but one egg and

this it places upon the point of a rock, with nothing like a nest or bed beneath it, yet so firmly that the mother can go and return without injury to it; but if anyone move it from its place, by no art can it be fixed or balanced again; left at liberty it straightway rolls off and falls into the sea. The place bedewed with a thick and viscid moisture which setting speedily, the egg is soldered as it were, or agglutinated to the subjacent rock.

Even better known is the assertion that the egg will spin on its own axis instead of rolling off the ledge. Usually, however, there are small crevices, or protruberances against which the egg can lodge, although some do get knocked off in hasty departures. Many coasters passing the Bass blow their hooters just to see the clouds of guillemots smoke out from the cliffs – a thoughtless practice which causes considerable loss of eggs.

The chick is downy from hatching and leaves its ledge long before it is fully grown, a hair-raising procedure. I witnessed it for the first unforgettable time at dusk, one grey, mild July evening at the base of the towering east cliffs. The clang and clamour of the gannetry was quietening, for gannets do not come in during darkness for fear of landing awkwardly or at the wrong nest. A few herring gulls still yodelled and 'klewed', as they would all through the short, summer night. Suddenly a wild, ringing call cut through the gloom. Not a whistle, nor yet a chirrup, it was a two-part call, rather like a short, violent, referee's whistle with a sharp break in it. It came again and again, with thrilling urgency. Among some guillemots, bobbing on the sullen, infinitely uninviting sea, I spotted a chick. It had thrown itself from a ledge above, though I hadn't seen it fall. Three highly excited adults attended it, constantly false-bathing and flipping beneath the surface. The chick had latched onto one, so closely that it seemed almost to have climbed onto its back and indeed, for a short time, may actually have done so. All three of them swam out to sea, eastwards. Guillemot fledging time is eerily exciting. Partly it is the dusk, partly those searching calls mingling with the rise and fall of the adult's gargling, partly the thought of those youngsters, unable to fly properly and yet hurling themselves from fearsome precipices, down to a sea that often enough sucks and gurgles among blackened boulders, or surges up the barnacle-encrusted base of the cliffs, whilst the predatory gulls wait on. How could the chick's own parent (for surely it would be improbable that every chick waited until both were in attendance) find it in the noise and excitement? But apparently it does. And if this seemed difficult on the Bass, what about the stupendous bird bazaars of the Arctic circle, the colossal cliffs of Greenland, Novaya Zemlya, Jan Mayen Island and Spitzbergen, with their millions of guillemots? It certainly helps if the parent can accompany its chick on this perilous first flight, and many do so, following its erratic course closely and landing with it. But many must lose contact in the mass fledgings which often occur, with hundreds of chicks in the air simultaneously. The young guillemot must afterwards be attended and fed at sea, for presumably it cannot cope alone in its half-grown state.

The kittiwake's fledging, from mid-July onwards, is far different, but equally fraught. Kittiwakes are cliff-adapted gulls. They are intensely territorial and, with impressive but occasionally misplaced faith in their building skills, stick their nests on the tiniest ledges and protuberances. This sort of site makes their first flight a life or death affair. Not for it the gangling incompetence of the juvenile herring gull, whose ugly mottled forms, still tatty with down, we saw leaping and jerking as, with excited squeaks and squawks, they practised becoming airborne. The young kittiwake must fly from its nest and return to it first time. If it flutters down to the sea it is in great danger from the large gulls and is unlikely ever to regain its nest. The sight of a herring gull, or a greater black-back, tugging at the floating corpse of a juvenile kittiwake is common at fledging time and often they do not wait for it to die. They set upon it and bludgeon and shake it to death. Immature greater black-backs regularly turned up at the Bass, as no doubt elsewhere, for the harvest. Gulls are great opportunists, always on the look-out. On one occasion I saw an adult kittiwake in the sea, in some distress after a fight with a rival. The loser was too exhausted to take off properly but managed to flap onto the rocks. Instantly, a herring gull, obviously reacting to its enfeebled state, was onto it, hammering it, and would no doubt have killed it but for my intervention. All that had been needed to trigger the gull's attack was that slight weakness in the kittiwake's movements.

Sometimes young herring gulls suffered fatal drubbings

Shags and kittiwakes.

from gannets when, unable to cope with down-draughts, they crash-landed amongst them. This sent the gannets berserk. Whether they have an inborn reaction to small, brown, potential predators such as rats or stoats, I do not know, but they simply worried the unfortunate gull to death. Yet, at sea, the same gannet attacked by a bonxie – which is simply a brown 'gull' – will usually surrender and regurgitate its food. Without the slightest doubt, an adult gannet is more than a match for a skua in a straight fight, but it doesn't fight. Why? The answer must be that the balance between the potential cost of fighting and the advantage gained by winning does not favour the latter. A slight injury could be serious whereas the loss of a couple of mackerel is not. In the colony, the situation is reversed; there, it may be the success of its breeding attempt – a substantial investment – that is at stake. Naturally, the gannet is not conscious of this calculation but then it doesn't have to be. Natural selection can work perfectly well without that.

The gannets share the rock with another Pelecaniform, the shag. There are no cormorants although the Lamb, a little further up the Forth, holds a colony. Neither of these birds form large colonies because they fish quite near to their breeding places and the food available in such a limited area could not, around Britain at least, support huge aggregations. This may partly account for the marked disinclination of these two species to nest on the same island, for they take some of the same food-fish. On the Bass Rock shags nest mainly in three roughly comparable groups, one at the base of the east cliffs near the caves, (40–70 nests), another (usually around 60) on Shag Rock (a flat-topped slab separated by a deep fissure from the Battlement slope) and the third (around 40 but up to 60) from the base of Cable Gulley round to Gannet's Graveyard, a lethal outcrop at the base of the west cliffs. Together with odd nests here and there the Bass 'shaggery' typically adds up to around 200 nests, though in some years substantially fewer.

By March, the bottle-green, emerald-eyed males had already grown their lovely, recurved crests. Dear, droll old shags with their spare bodies, huge feet, serpentine necks, sepulchral croaks and perpetually anxious mien. Their displays are comical, too. The first time I ever saw a male darting its head this way and that and then throwing it back

114

A puffin, probably a pre-breeder.

towards its tail, I thought it was striving to nail a particularly bothersome fly! I even (now to my embarrassment) described it as such in my diary. In fact he was inviting a nearby female to join him.

By early July Shag Rock usually held many nests with almost fully grown chicks, two or three drab brown versions of their parents. Yet they had not laid earlier than the gannets and the latter's single chick was still less than half-grown. And that, despite the gannet's phenomenally rapid growth when compared with other members of its own family. The shag chicks grew faster because their parents fished inshore and thus fed their young more frequently. By mid-August there were more than 200 youngsters swimming off the rock, practising their fishing and still receiving food from their parents, before dispersing. By then, the season's crop of young kittiwakes, guillemots, puffins and razorbills had all gone and only the gannets and fulmars, together with a few late gulls, still had chicks to feed. This end to the main seabird season always came surprisingly abruptly. In July things seemed at their height, the emptiness of winter a faint memory. Shortly afterwards there were gaps, silences, intimations of autumn. The guillemot ledges were empty, the part-grown chicks away at sea. Even before the black-collared young kittiwakes had all quitted their fouled pads the tempo of their parents' activity had slackened and there were signs of the early spring land-shyness. Maybe it is the same with us. We grow suddenly old and wonder where the summer days all went.

9

Fruition

But above all, writing was most astonishing to them:
they cannot conceive how it is possible for any mortal
to express the conceptions of his mind in such black
characters upon white paper.
Martin Martin describing the St Kildans

We believe a scientist because he can substantiate his
remarks, not because he is eloquent and forcible in
his enunciation. In fact, we distrust him when he
seems to be influencing us by his manner.
Peter Medawar, *Pluto's Republic*

Autumn was the time for migrants. After easterly winds,
misty days in October saw the Bass playing awkward host to
hundreds of temporarily stranded visitors: warblers, flycatchers, redstarts, siskins, goldcrests, redwings, blackbirds, robins, woodpeckers, treecreepers, owls, falcons and
many more. They did their best to forage but it was hopeless.
Siskins swinging on thistles, treecreepers climbing the
chapel doorway like their cousins the wallcreepers, redwings
poking around in the gulleys, blackbirds flinging gannet nest
material aside as they do woodland leaf-litter and woodpeckers fruitlessly hammering the mallow stems. Under better conditions, in September, some of the flycatchers and
robins even set up temporary territories, chasing away intruders and displaying threat. Similarly, the migrant meadow
pipits were chivvied by our own rock pipits. Two Lapland
buntings fed for a day or two by the pool on top of the Rock,
running, usually horizontally, through the tangle, lovely
little birds. For us, they were all delightful visitors, shoulder-
ing aside our old faithfuls and adding a touch of glamour.
The biggest fall of robins was in spring 1962 when, on 18
March, there were well over a hundred on the Rock, together

with bramblings, Greenland wheatears and woodcock. Occasionally we were visited by small parties of wandering crossbills. None of these birds were far out of the ordinary, but they added a touch of excitement.

On 11 October 1962, a day of surging grey sea and close-wrapping mist, we left the Bass. By dawn, flocks of two to three hundred redwings strung out as they flew past, calling thinly. Fieldfares, blackbirds, thrushes, ring-ousels, robins, bramblings, siskins, chaffinches, redstarts and meadow pipits thronged the Rock. The air was alive with calls and flitting shapes and we hated the very thought of returning to the bustle of Oxford for the winter. How insular one becomes. But the chapel on the Bass would still be there to welcome us back in the spring of our final year, 1963.

Accommodation in the winters presented a problem. Niko Tinbergen's Behaviour Unit had graduated from its row of hen huts on the roof of the Zoology Department (I could just about turn round in mine) and now operated from Number 13, Bevington Road, a gaunt terraced house. True to form, Niko had let it be known that Bevington Road was there to foster enthusiasm and discussion among ethologists. It certainly fizzed with activity and many a wandering student bedded down for a few nights there. Juan Delius nurtured a gaggle of sulky-looking herring gulls in the backyard; Richard Dawkins, author of *The Selfish Gene,* was experimenting on the acquisition of feeding behaviour in chicks; Hans Kruuk and Ian Patterson shared a room when not chasing black-headed gulls. In an attic, Dick Brown brooded over rows of bottles containing dancing fruit flies. In the basement, Esther Cullen's tanks of sticklebacks were convincingly demonstrating that ritualised behaviour patterns are inborn. Every lunchtime Mike Cullen, his cerebral hemispheres encased in an outsized helmet, spotted like a ladybird, buzzed up on his dilapidated scooter to spend equal amounts of time and energy consuming jam and bread, his staple diet, and dispensing valuable advice and criticism. We had brought our paraffin oven off the Bass and, just for a day or two, set it up in the corridor, because we hadn't a room. Professor Pringle, who happened to visit the place just as we were frying lamb chops, courteously reminded us that we were no longer a law unto ourselves. He was right, and that was one reason I liked the Bass.

That winter was a lucky one, for in February I went up to Aberdeen University to talk about gannets – a trip which shaped the rest of my life. Aberdeen's Department of Natural History has been headed by many famous zoologists – Thompson, Ritchie, Hogben and Alister Hardy among them – but pride of place must surely go to the incumbent of that time, Vero Wynne-Edwards. He had just written his remarkable book on the regulation of animal populations and, although its central tenet has since been vigorously attacked, it is universally acknowledged to have been an epochal contribution. Seabirds, gannets and fulmars especially, have always been a great love of his, and a few years later he invited me to Aberdeen, where I have remained ever since. Gannets and fulmars have both figured prominently in the work of the department and thrown some light on the nature of recruitment in seabirds – a prime Wynne-Edwardian interest. It was a grand old department in Marischal College, steeped in the Natural Historical tradition of its great professors. I say 'was' because, in the last decade, it has more than doubled in size and moved from the ornate, silver-granite Gothic of Marischal College into an enormously ugly concrete box, part of the rash of monstrosities that we suffered in the 'sixties and early 'seventies. Even on its own expensive terms it didn't work. All the heat went to one side of the building leaving the other like an ice-box. And now the concrete is cracking.

A department, or any equivalent 'hunting group' has an optimal size, a magic number, below which it is ineffective and above which it begins to fragment and generate conflict. A group of about a dozen, give or take, seems to be big enough to benefit from group experience, strength and diversity yet small enough to foster intimate co-operation. Such a group rallies round a leader instead of throwing up competitors. Aberdeen, alas, has long since passed the magic number.

At this time, Adam Watson took me on to his red grouse moor where a study of social behaviour, which would have gladdened Fraser Darling's heart, was under way. On this brilliant morning in February a cock was already displaying to a neat hen, fanning his tail turned sideways to the female and snapping his head from side to side, red wattles gleaming in the sun. He wasn't, as one might have thought, establishing his territory. This he had done the previous autumn and now he was merely reclaiming it. Most of the birds which had been excluded in the autumn – more than half the population – had already died, having been forced into marginal habitat. They had been 'surplus to requirement', although nobody had suspected that there were so many of them. What did it matter if a few were taken by predators? Where, now, even in the keeper's grouse-orientated framework, was the case for killing off raptors that did no harm? As Adam bucketed us across the rough moorland in his Land Rover, looking for a few grouse, I blessed my good fortune in having hundreds of displaying birds to watch whereas Adam had to patrol the moors for hours to see one or two displays. But it would be mistaken, simply on account of having spectacular displays, to think that the structure of seabird societies is easier to understand than that of grouse. The displays themselves are common and distinctive – I could collect a thousand instances of the gannet's territorial display in two or three days whereas Adam might take years to garner a thousand of the display I had just seen. But the functional aspects of the grouse's social system as a whole may nevertheless be easier to investigate than that of the gannet. The autumn exclusion of surplus males, for instance, was readily apparent – a fascinating social system which could be demonstrated without much ethological study.

Birds differ greatly in their communication behaviour. The gannet has an intense and complex greeting ceremony, performed with undiminished vigour hundreds of times each season, year after year. The golden eagle has none. Why? Again, as I have already mentioned, the fulmar which, like the gannet, is both a seabird and colonial, has only a simple head-swaying and cackling whereas a herring gull employs a whole repertoire of distinctive displays. Both of them have to stake out a territory, attract a mate, maintain a pair-bond and co-operate during incubation and rearing the young, yet one species uses a battery of displays and the other makes do with one.

Sometimes, limited answers do seem possible. In the Galapagos, for instance, the great frigatebird has hardly any territorial display, in marked contrast to the gulls and boobies. It doesn't claim a nest-site and then try to attract a female to it. Instead, it joins a group of displaying males in an

A male great frigatebird displaying its scarlet throat sac.

attempt to lure down the females who soar above. If it is
unsuccessful, it may move elsewhere and join another display
group. Communal display such as this is hardly compatible
with fierce territorial antagonism. Indeed, the males are often
in bodily contact – an unthinkable situation for a booby or
gannet. And even if a private display site *were* desirable
within the group, it would hardly be worth fighting and
displaying in defence of a spot that was quite likely to be
vacated anyway. And although the display by which the
males attract the females is dramatic enough, the pair have
no subsequent greeting ceremony and when, later in the
season they meet at the site they simply ignore each other.

121

The free-flying young frigate is fed by its parents for up to a year after fledging. Even afterwards, many starve when forced to forage for themselves.

When gannets re-unite on the site they perform an ecstatic mutual display. But whereas gannets pair for life, frigates, as we now know, dissolve their bond after each breeding attempt. So, on this restricted level, one can understand why frigates behave in one way and gannets in another. But there is always another 'why' behind the first.

Initially, the frigates were a great puzzle in many ways. Eventually we discovered that they couldn't have permanent sites and faithful pairs because they take much longer than a year to produce an independent youngster. Whilst they are still busy with their free-flying juvenile, a year after they started that particular breeding cycle, another pair may take over the site to begin *their* attempt. And eventually, when they do stop feeding their large offspring, they go off to sea to moult and recuperate. But as they live in the tropics where there are no definite seasons to act as 'timers', they can hardly be expected to synchronise their return, in addition to which, they have no fixed site to return to. So they are most unlikely to find each other again, which may be partly why they have not evolved a pair-bonding display. The gannet works under a completely different set of circumstances.

'Explanations' like this are not, and cannot be, capable of proof. They are merely correlations and more or less plausible interpretations, tying the facts together in a way that makes sense. The evolutionist believes that behaviour, even human behaviour, always makes sense, even when it appears not to. By the time we started our last season the gannet story was beginning to make sense.

It was 25 March 1963 when we again set foot on the Bass. The gales and hardships of the previous two years had been forgotten. We remembered only the sea-girt Rock and its seafowl. The hut soon lost its forlorn air and once more became cheerful and busy, the copper urn steaming on the little black oven, tea brewing and there was the familiar squabble for table space between books, papers, cooking gear and food. Who wouldn't have traded Oxford's dreaming spires? Samuel Johnson, no doubt; he who described the rugged Buchan coast at the sea-cave known as the Bullars of Buchan, as his idea of hell. Poor Samuel. His celibacy, or something, cost him dear, for he was rarely at peace with himself and, no doubt, dreaded the solitude of wild places.

This was to be an especially busy year. In addition to another season's fieldwork I had to write my DPhil thesis and organise our forthcoming trip to the Galapagos where we hoped to study some of the gannet's tropical relatives, the boobies. How we were to fit it all into one summer I hardly knew, but at least we would be spared the distractions of Oxford. As an earnest of my good intentions I immediately arranged my files, books and card index on the shelves at the back of the hut. I felt sure I would write better here, where I could step outside and be among my birds, than incarcerated in Oxford. In fact, despite the lack of facilities I was so keen to return that nothing would have deterred me. Writing is a matter of motivation, and on the Bass I was motivated.

123

Others raised doubts, but they didn't know what it would have cost me to pine for the Bass. It was an adventure to tap away on the typewriter in an old ruined chapel on the finest seabird rock in Britain, a chore to do the same thing in a city room. I feel almost ashamed to recall the eagerness, almost like an escaping prisoner, with which I turned the Land Rover Bass-wards. On that first day back, how grand it was to see again the shabby little coasters butting through the channel beneath the frowning battlements, dipping their rusty bows into the chop, with a thin following of hopeful gulls. In no time the hut was in splendid order, all our gear manhandled up the hill, a dozen letters written, the rubbish burned and the lavatory tent erected.

Two weeks later the sea-scurvy was emerging, the first gannet eggs appearing and spring was on the way. More than that, the American Chapman Foundation wrote to tell me that I had been granted five thousand dollars to carry out research on tropical boobies. Although I am sure Frank Chapman would have approved, I could hardly believe our good fortune. I still found it hard to believe that anybody would give money to enable me to watch seabirds on a desert island. What a stimulus to complete the gannet work and leave with the job well done. But what a temptation to dream of tropical islands inhabited by red-footed boobies, frigate-birds and tropicbirds. How would they behave, compared with my gannets? 'In verba nullius' – on the report of none – we would see for ourselves. The Royal Society's motto reminds me that my old professor, H. G. Callan of lampbrush chromosome fame, had been elected to their august ranks. Give me gannets any day, though I don't expect them ever to land me in the sacred precincts of Carlton Terrace. Still, our revered Niko has been elected, another mark of respect from the scientific establishment. Many people think that only laboratory work is real science. The word is much abused anyway and most people don't really know what it means. They think of it as a stamp of authenticity whereas it is merely a disciplined and objective approach. There is no such thing as 'scientific truth' – just 'truth'. Karl Popper's falsifiability criterion is not an adequate definition.

I wrote immediately to the great American seabird bio-logist, Robert Cushman Murphy, tall, handsome and distin-guished – a man who travelled through life 'first-class' (I tend to go stowage). He was an authority on the Peruvian Seabird Islands, whose immense colonies of boobies, cormorants and pelicans beggar belief and which I was determined to see for myself. When, eventually, we did clamber up the corroded iron ladder against the rocky flank of Guañape Norte the sight was sufficiently stupendous. I wrote, also, to an expert on solar stills in case we found ourselves attracted to a waterless and uninhabited island for a year. In the event, we did exactly that. Du Pont helpfully offered to provide some mylar (a tough, transparent material) free of charge. It all put flesh onto the approaching adventure. I could imagine the still; what I could not imagine were the surroundings, but of that more later.

Three weeks later Fred brought across another exciting batch of mail. Although at the time it seemed the height of indulgence (almost a fifth of our total year's income) I bought a 400 mm Novaflex lens with pistol grip and follow-focus. It has been, and is, invaluable. Its enamelled barrel, once pristine, is now chipped and scratched; seaspray has corroded it and the lovely blue bloom on the lens has disintegrated but it has mirrored rare and wonderful scenes. I couldn't afford the leather case, so Bertie (one of the keepers) made me a wooden one. Alas, the camera behind the lens, an ancient Exakta, caused endless trouble, wrecking irreplaceable sequences of albatross display and much else. I have always felt too poor to buy expensive equipment but it has been misplaced prudence. It would have paid to go for first-class cameras and binoculars. Maybe my old grandfather's hand reached through the years, for he was the strictest of budget balancers and never borrowed a farthing. My mother's creed, bless her, was equally simple – if you can't afford it, don't buy it.

Then came an incredible letter from Niko, saying that he had been invited to visit the Galapagos with a team of scientists to celebrate the opening of the Charles Darwin Research Station in 1964. Several eminent naturalists were to attend the ceremony and remain for a few weeks to initiate research projects. Could he stay with us on Tower? Could he! Who better? Delightful as a companion, a naturalist to his fingertips, he would revel in the wildlife of Tower Island and provide tremendous stimulus.

Those were heady days, with never a dull moment. Silly as

125

it now seems, I added to my confusion of activities by writing a paper for *IBIS* on my twinning results. Surely that could have waited until the thesis was out of the way. There was one unpleasant hiatus. My smallpox injection threw me into a raging fever, alternately sweating and shivering. I lay on my campbed with two hot water bottles and piles of blankets, teeth chattering, babbling about gannet graphs. It was late June and dense mist had closed over the Rock, grey, miserable and damp. The foghorn boomed sepulchrally and the compressed air hissed in the pipeline. That fog and my misery lasted for six days, stolen from the busiest period of my life.

By early September little more than a week remained before the thesis, the results of our labour, had to be posted to Oxford. Half of it was still to type and some of it still to write. It was a continuous fight against time, all-night work. For a moment, fieldwork was abandoned. The hide stood empty. A great gale blew up just when the typing for all three copies was neatly piled on the shelves of the hut. In my overwrought condition and the pitch darkness perhaps the din sounded worse than it was, but I was so convinced that the roof would lift and my work be scattered upon the uncaring waters of the Forth that I lay spreadeagled on top of it all night. Morning found it creased but safe.

The great day arrived on 16 September. The thesis was to be despatched by boat to North Berwick, there to be posted by Fred to Oxford in time to allow 24 hours for binding before meeting the deadline for handing in. It was a near thing. The tide was ebbing and Fred, hanging on at the Bass landing place, was in danger of being stranded outside North Berwick harbour. I actually tied the package as I ran down the steps from the chapel. This ridiculous pantomime was unavoidable because when Fred arrived to pick it up June still had the list of Contents to type and they hadn't even been roughed out; I dictated them. We couldn't have risen earlier because we hadn't been to bed! All this was the fruit of an unrealistic deadline, self-imposed. But there was a year's expedition to an uninhabited island still to organise before October. I naively believed that, having accepted one of the set dates for handing-in, all my toil would be wasted if I missed Fred's boat by a few paltry minutes, for that would have left no chance of catching the post and meeting the

126

deadline. Now, I know that under the circumstances it would have been accepted a few days late. But the relief as I climbed wearily back up from the landing was quite unreal – one of the highspots of my life.

A few days afterwards we had a little party in the old chapel. Fred anchored the *Girl Pat* just off the gulley where she lay, rocking gently, cheerfully lit inside. Melon, ham, mushrooms, tomatoes, beans, potatoes, fresh peaches and cream, cider, biscuits and cheese and coffee. It went down well and my wonderful wife notched up yet another 'first'. It was low tide at midnight as we escorted Fred, Joan and their daughter Pat to the landing. Sea-slaters swarmed everywhere and the massive tangle of glistening kelp rose and fell sluggishly.

It was a happy time – perhaps uniquely so. There we were, in our secluded little world, three memorable years behind us and the great Galapagos adventure ahead. We envied nobody and I had not the remotest inkling of the distant storm clouds. The security-minded might have questioned the wisdom of postponing job-hunting, but we were not security-minded. The weather remained fine and we were free to enjoy it. The rock was crawling with actors in period costume, for the BBC were shooting scenes for *Kidnapped*. We watched briefly from the battlements before turning to more interesting things, chiefly the manufacture of stout wooden boxes to carry our gear to the Galapagos. We made them from our hut shelves – a mournful task, dismantling our home. Soon it would disappear and the chapel would return to its empty brooding, overrun with nettles and mallows. On wild, dark nights the cold wind would finger the old stones, searching for weakness, and there would be no cheerful light shining through the portals. Our sojourn, which to us seemed a lifetime, was but a flicker in the Bass centuries. When the last stick had been removed and all was tidy, I wandered up by myself, just to say goodbye, to put the moment quietly into its place. The chapel and the moment are still there.

It seemed right that the hut was not allowed to remain. Once there is accommodation on an island, the whole game changes. It is better as it is. The Bass is the only place with which I have been deeply involved that has not changed for the worse. The 'sixties and 'seventies wrought unbelievable

127

environmental havoc, world-wide, and the 'eighties may be worse still. Take for instance, merely the places I have known. Christmas Island was largely pristine rainforest, an absolute gem. Now it has been ravaged by phosphate mining, large areas reduced to spoil heaps and naked pinnacles. Then there is Azraq, a rare jewel of an oasis in the Jordanian desert. Its clear pools spilled over into acres of permanent marsh, alive with birds. Instead of land surrounded by water, it was water surrounded by desert, as inhospitable to land-birds as the sea. A crossroads for the immense flow of Palae-arctic migrants, it is one of the finest bird-watching places in the world. Azraq, too, has been spoiled. Its water has been led away in every-increasing volumes. The marsh may be slowly dying. A highway has been driven through the heart of the oasis. Even Aberdeen, the granite city, capital of the distinctive Buchan area, for centuries a dignified and un-changing centre of farming and fishing, the academic outpost of the north to which Scotland sends so many of her sons from far-flung isles and glens – Aberdeen has changed. It is now an oil-dominated city, surrounded by huge industrial estates, its outlying villages swamped by housing estates, its old social structure tottering. More money, more crime, more flashiness, more restaurants, warehouses, office blocks, gargantuan lorries, boutiques and strippers. Less of old Aberdeen. Developments can be so insensitive, sometimes downright crooked. Take, for instance, the Loch of Strath-beg. Strathbeg is a very special brackish loch nestling behind high dunes to the north of Aberdeen. It was claimed that North Sea gas just had to be piped ashore right through, or under, the loch. Of course it was regrettable, and naturally alternative routes had been carefully explored. But there it was – just one of those unfortunate things. The Loch would have to be drained. How many Strathbegs have been need-lessly destroyed, for in the event, and for quite other reasons than to accommodate tiresome conservationists, the pipeline was re-routed a paltry three miles to the South – quite easily and perfectly feasible in the first place. Strathbeg remains a marvellous sanctuary.

10

Castaways

The thrill of one's first desert island is a quite inde-
scribable thing and the fascination of these barren
Galápagos is inexplicable. A land pitted by countless
craters and heaped into myriad mounds of frag-
mented rock: gnarled trees bleached by salt-spray,
with twisted, stunted branches; grotesque cactus
stiffly outlined against black lava boulders, – this
picture, truthful in its bare statement of facts, cannot
convey the feeling of mystery that led us on and on,
wondering always what lay just beyond the next barrier of sombre rock.

　　Beebe, *Galapagos: World's end*

From the basalt of the Bass to the lava crust of Tower – Quita
Suena – Devil's Island. From gannets to boobies, herring
gulls to lava gulls, hut life to tent life. The very same oven
and refrigerator, packed away on the Bass, was unpacked on
Tower.

　We pitched our tent on a tiny coral beach at the head of
Darwin Bay, a drowned volcanic crater. On one side were
low, slab-like cliffs fringed with dried, silvery shrub. Behind
stretched a crystal sea-pool straight from a fairy tale – a
shallow lagoon backed by low green cryptocarpus shrubs in
which nested red-footed boobies and great frigatebirds. On
the other side a tidal creek wound between cactus-fringed
cliffs and past a great sweep of cryptocarpus dotted with
clearings in which nested masked boobies. If we lifted the
back flap of the tent we could observe the frigates and boobies
as though from a rather distant hide. From the verandah the
whole of Darwin Bay spread out before us, and it was never
empty of birds, sea-lions, sharks or dolphins. And the sun
shone with tropical warmth. After three years on the Bass, it

was going to take a lot of sun to satiate us. Already, the Bass seemed a distant memory – we had been travelling so long.

Two days out from Southampton we ran into a full-scale hurricane. Spray cleared the funnels in sheets. The *Queen Elizabeth's* great bow dipped ponderously into the troughs only to rise majestically beyond the horizon and up into the sky. The seas towered monstrously, far higher than the top decks when her bows were down. When they were up, the view was of deep valleys and slopes, from which sheets of spray curled and smoked. Yet, there were frail kittiwakes hugging the contours, a few grams of feathers riding out a hurricane. In mid-Atlantic we ran into several hundreds all flying with the wind. Among them were two juveniles. Maybe they were brothers, now far from their noisy colony. They were accompanied by several fulmars, a single guillemot and three shearwaters. About 100 miles east of the North American coast we encountered several groups of gannets. Canadian gannets are exactly like their British counterparts in appearance, ecology and behaviour and there may even be some slight interchange between the two populations. Certainly their oceanic range could just overlap.

The boxes, our precious boxes, lighthouse green, neatly numbered and with typed content-lists pinned inside the lids, were transferred from Cunard to Grace Line and two days later we set sail for Ecuador. It was a slow and tortuous journey from our Newark berth, under Statten Bridge and past the Statue of Liberty and Manhatten. What a waste it is to travel by air and what a loss we have suffered in the demise of the great shipping lines. Nowadays we sit in silver tubes and are transported joylessly through the stratosphere – 28,000 feet, flying time 16 hours 20 minutes, local weather fine but cloudy, please remember your hand baggage, thank you for flying with us and we hope to see you again. But in 1963 we had never flown. Now we were skirting Haiti and our first masked booby came alongside, very gannet-like with his dazzling white plumage. They looked lighter and more buoyant, which of course they are. Although almost as large as a gannet, they weigh only half as much. They are tropical seabirds and don't need the thick insulating layers of fat, nor the weight to dive so deeply and hold such strong prey. At Barranquilla in Colombia the river was in flood, sweeping thousands of tons of precious top soil away from

130

The Galapagos Islands, of which Tower Island and
Hood Island form part.

the interior to deaden the sea – a double death. Cattle egrets
and great white herons perched on masses of floating vegetation and thirty or forty turkey vultures with huge, uptilted
wings and red heads soared above. Ospreys plunged into the
swirling brown water and swallows hawked above it – not
our familiar birds but velvet green and rough-winged swallows. None of these birds inhabit the jumbo's stratosphere.

We crossed the equator on 2 December and Neptune was
in severe mood, plastering me with raw eggs, tomato ketchup
and whitewash. I added red-footed boobies to my list of
sulids – a long-winged, agile booby, the male barely a third
as heavy as the gannet. Later, their tinny and vibrant calls
sounded day and night around our tent and to this day I
cannot hear a football rattle without seeing red-feet. Soon,
Guayaquil – 'hell-hole of the Pacific' – hove into view and,
with it, a pronounced sinking feeling. I had been putting off

131

the moment when the cocooned life aboard ship would end and we would be precipitated, green as Bass grass, into all the tensions and turmoil of the next phase. North Berwick, where art thou? Somewhere in the *Santa Mariana's* hold were eleven vital wooden boxes. Would they be put into bond, disappearing into the dark recesses of some monstrous custom shed, never to emerge again? What if the *Cristobal Carrier*, the wreck which plied between Guayaquil and 'Las Islas Encantadas', the Galapagos, was almost due to leave (useless to ask about this in advance) and would do so before we could extricate our boxes? What if we were charged huge sums for each day's storage and the *Carrier* was held up for weeks, as sometimes happened; her sailings were notoriously erratic. We were on such a shoestring – six hundred pounds for the two of us for a year, to cover everything except our major travelling – that our entire budget could be wrecked.

But somebody was looking after us. Roberto Gilbert, a mercurial Ecuadorian surgeon who had befriended us on board had, by chance, removed an appendix from the Chief Customs Inspector at Guayaquil. The same officer might, who knows, one day need heart surgery, a cataract operation or a bunion removed. Roberto, in his private Clinica Guayaquil, would be the man for the job. The upshot was that we sped through Customs without a box being opened and, moreover, found storage space on the roof of the Clinica. This was crucial; our equipment was eminently stealable, yet it had been impossible to arrange storage in advance. We stored ourselves in the cheap but clean Pension Helbig, three meals for little more than a pound a day.

Once John Lacey, the British Consul, had decided that we were basically sane rather than a couple of hippies, he abandoned his stuffy office for the dim delights of Elias Mayorga's Chandlery. For the equivalent of around £300 this swarthy gentleman undertook to supply and crate food for a year. Bananas were cheapest. A reject stem weighing 50 or 60 pounds sold for a few pence. The flat guayas plain produced unbelievable mountains of bananas. The plantation workers lived in miserable huts beneath dripping banana leaves, humid and enclosed, walking everywhere on slippery red earth between regiments of the giant monocotyledons, a lifetime of banana monotony. I shudder to think of it.

Dr Gilbert was undoubtedly a medical whizz-kid. At 46,

he had just paid off the mortgage on his clinic. The place and its equipment was now his, at today's prices worth far more than a million pounds. It was the best hospital in Ecuador, beautifully built of marble with departments of radiology, endocrinology and dentistry and fully equipped with kidney machines, lung-unit, low-temperature gear, and much else, all of which Roberto treated with the utmost sang-froid. He had performed the first mitral valve (heart) operation in Equador and he actually sewed back a sailor's completely severed hand, though it was eventually rejected.

Enthusiasm was his life-style and soon, not content with lists and advice, he was robbing his storerooms and cupboards of expensive drugs for us. At one o'clock in the morning, still in surgeon's gown, he bundled a wretched intern into the back of his pick-up 'for some fresh air' and we all roared off on a tour of Guayaquil. Surgeon he may have been; driver he was not. He staggered along in top gear at 10 m.p.h. and then, to make up for it, began a perfect frenzy of gear changes, all ill-timed. He wove a crazy route, stopping, turning, stalling and without the least warning, executing split-second changes of direction. The common people of Guayaquil loved him. Everybody seemed to know him, though he was hard to miss as he strode through the streets in his surgeon's gown in quest of some potty little article that he thought we should have – like a shotgun. Why he took so much trouble over us I never knew. Possibly it was June's magic. In those days I never suspected such things.

Sailing day was 11 December. Down at the old *Cristobal Carrier* the scene, as they say, was animated, not to say agitated. Forty-seven boxes of Elias Mayorga's provisions were ticked aboard (why only forty-six came off at the other end I never knew but whoever picked the forty-seventh got the only one with chicken and ham; we had to make do with tuna for a year). For the big stuff, though, we needed a lorry and a gang of labourers to transport our Bass boxes to the *Carrier*. Those Ecuadorians, of indeterminate age and slight though sinewy build, performed prodigies of strength as they each carried a devilishly heavy and awkward wooden chest up a slippery gangplank which was not only sharply inclined but also tilted to one side. There was a steady stream of cargo going aboard at the same time and in the middle of all this a second lorry arrived with an old rowing boat made

133

of ironwood (and weighing the earth), which I had bought for £12 up river. That had to go on deck along with a few tree-trunks and some banana stems.

Dr Gilbert still hadn't quite finished with us. Returning, exhausted, to the Clinic we were grabbed and together with his radiologist, an ex-colonel, taken off to the Ministry of Defence. There we bumped into the Minister himself just off to the Town Hall Conference Chamber, whereupon our colonel dictated a letter requesting all necessary Civil and Military assistance for 'Her Britannic Majesty's Special Envoy, Señor Doctor Bryan Nelson' and, following the Minister to the Town Hall, penetrated a succession of armed guards and obtained the Minister's signature whilst he was actually in Conference. Can you imagine the British Minister of Defence responding thus on behalf of some utterly insignificant Ecuadorian? I found it all rather bewildering.

The *Carrier* sailed at about 11 p.m. and it was a relief to slip quietly down the muddy river Guayas in the velvety darkness, though the boat inspired little confidence. She was basically a flat-bottomed landing barge with a battered old cricket pavilion on top and she rolled hideously, but the sea was calm and mercifully remained so. After this, only one more boat and we would be there. How far from the reality we had been during our planning on the Bass Rock. Perhaps luckily so, our imaginations had been fallible guides. Meanwhile we relaxed and watched the birds. Waved albatrosses soon appeared and then storm petrels (Madeiran) began to come aboard at night, attracted by the lights. About 150 miles east of San Cristobal the sea was alive with them, in twos and threes with a strong, wheeling flight punctuated by slow, butterfly strokes. Then came the first sightings of the lovely Galapagos swallow-tailed gull. Flying fish leapt clear of the sparkling sea and a large shark cruised near the boat. We were to become familiar with sharks, which in the Galapagos are harmless. Nobody takes any notice of them and we never heard of anybody being attacked.

Three days after leaving Guayaquil the *Carrier* nosed into Wreck Bay, blowing three raucous blasts. The gathering light showed an attractive settlement fringing a pleasant beach with cloud-capped hills behind. The painted wooden houses, well-bleached by the sun, sat right on the beach without tarmac, road or traffic. At that time Wreck Bay held

Swallow-tailed gulls.

745 people out of San Cristobal's 1,399. Santa Cruz came next with 609, then Isobella's 283 followed by Floreana, 46 and San Diego, 36. The other islands were uninhabited.

On we sailed to Santa Cruz and Academy Bay, where the Charles Darwin Research Station was just about to be opened. We off-loaded our mountainous heap onto the station raft, which rolled alarmingly, threatening to slide our life-supporting boxes overboard. Santa Cruz was dominated by the Angermeyers, a German family, and there were still several of the band of Norwegian settlers who arrived in the 1920s. Academy Bay is a honey-pot for visitors, one of whom was David Balfour in his yacht *Lucent*. This was to take us on the final leg to Tower Island. Before that, though, we went with David Snow (whose blackbirds I had inherited in Oxford) to Plaza and Daphne, where he was studying tropic-birds. Plaza is a low island, studded with cacti and succulents, much favoured by land iguanas and crawling with

135

sea-lions, turtles, boobies, tropicbirds, pelicans, frigate-birds, Audubon's shearwaters and swallow-tailed gulls. The omens seemed good! From Plaza we sailed to anchor near Baltra, in sight of Daphne Island. The creak of block and tackle, the slap of water, the splash of cavorting porpoises and the calls of seabirds served only to emphasise the peace. Then supper.

The next day was Daphne Day, the day of the blue-footed booby. Daphne is an extinct volcano. The steep flanks provide holes for tropicbirds and the flat-bottomed crater, in two tiers, is booby domain. The crater walls rise steeply from the flat bottom where the blue-footed boobies concentrate in the blistering interior, scattered on the ashy floor like currants on a floury plate. The blue-feet form a pure colony without a single pair of masked boobies, though these nest around the rim. It was a puzzling choice of an uncomfortable place, a broiling pit apparently without benefit of ocean breeze. The poor old boobies stood on their heels so as to raise the underside of the web above the hot ash which was indeed far too hot for our bare feet. But maybe the boobies were not so dumb, for Beebe says

> it was surprising on both days of my visit to Daphne to find the crater so cool. Although shut in completely on all sides, yet there was actually more breeze here than outside, the heated air from the white sand evidently ascending while cool, fresh air siphoned in from the notch on the eastern rim.

Nevertheless I pitied the chicks tied to their stifling patch in their warm down. But the nights are cold and even if they were not, a chick has to grow its down sometime. Temperature control is a problem for all seabirds and they meet it in different ways. Where food is plentiful one parent can afford to brood or shade its chick but where it is scarce, as in the Galapagos, it is often impracticable to do so and the young are left to their own devices as soon as they can cope. On Daphne, many small young were unattended and suffering harshly from the heat.

In the upper crater the young, all much of an age, were considerably larger than those in the lower. This state of affairs, which could never have occurred among the gannets of the Bass, immediately caught my eye. Here on Daphne, whatever controlled the time at which eggs were laid clearly allowed the birds a great deal more latitude for if it had been

136

important for them all to fledge by a certain date, there would not have been such a spread of laying. In the Galapagos the seasons are ill-defined, if indeed there are any. There is nothing comparable to the onset of the autumnal gales of the high northern latitudes nor the seasonal flush of shoaling fish of which gannets take such advantage. Indeed, as we were to discover, food frequently disappeared abruptly and unpredictably, so from that viewpoint there was no 'best' time to rear young. This phenomenon, which constantly recurred, seemed to me to exert a decisive influence on the breeding strategies of the seabirds.

Between Plaza and Santa Cruz we passed through a shoal of yellow-tailed mullet. The sea was coloured by them, its surface boiling as they threshed about, presumably feeding on a shoal of fry or perhaps a patch of invertebrates such as the marine pond-skaters which can turn up in mid-ocean. The Galapagos brown pelicans had gathered and were diving repeatedly. Grey phalaropes by the thousand surrounded the *Lucent*.

Every Galapagos island has its own climate, landscape and vegetation. We were to spend our year on the sharp and brittle lava of Tower and the time-smoothed boulders of Hood but, had we chosen to do so, we could have lived in luxuriant rain-forest or amidst lush upland pasture. On the high ground of Santa Cruz a Norwegian family, the Hornemanns, had made their home. Following in David Lack's 1939 footsteps, we went to pay our respects. The five-mile track from Academy Bay, always narrow and bouldery, became slippery with mud in the higher reaches. At intervals, the bleached bones of horses and donkeys marked the way. We soon learned that *all* visiting scientists wend their way to Mrs Hornemann's hospitable farm – Lack, Bowman, Kramer, Curio, Eibl-Eibesfeldt and, after us, Peter Scott and many more. Her husband, a brilliant chess player, but latterly a paranoid schizophrenic, had just gone for treatment to Guayaquil. In the humidity and heat of upland Santa Cruz even a walking stick takes root before it can be pulled out of the ground. She grew coffee, papaya, marrow, sugar cane, potatoes, beans, bananas, corn, lettuce, tree-tomatoes, tobacco, mangoes, avocadoes, a sort of turnip and no doubt much else. And she kept pigs, cattle, donkeys, horses and poultry. The house was saturated with the delicious aroma of

137

freshly roasted and ground coffee, the best I ever tasted.

The next day Sigvert Hornemann took us up to tortoise country, a round trip of a dozen miles. It was a tedious struggle through mud and sodden vegetation and over slimy, lichen-hung boughs. The fine drizzle turned into solid rain and some nettle-like plant gave us an irritating itch. The scratches garnered on this trip later festered and caused much trouble. I had to keep them dry, which is next door to impossible in the work we were doing and at first, on Tower, June had to carry me and all my equipment across every water barrier. Septic sores can be depressingly persistent in the tropics, spreading at an alarming rate. Eventually we emerged into open country at the foot of Table Mountain and there in front of us were two small ponds, covered with rosy sphagnum. Fourteen demure Galapagos pintails were displaying on them and nearby were nine giant tortoises, the largest weighing perhaps 200 pounds. We, in turn, were scrutinised by a one-horned bull, and wild pigs and goats were in evidence. Introduced animals have been the scourge of many oceanic islands and the Galapagos have suffered as much as any. Goats wreak havoc on the vegetation, pigs trample and eat tortoise eggs and young, and rats have brought some petrel colonies to the verge of extinction.

But these pleasant trips soon ended and we started loading up the *Lucent*. Twenty-five gallons of paraffin and sixty gallons of water went aboard. The sacks of flour, sugar and rice were vulnerable to pests but seemed to survive in the station warehouse. The boxes were ferried out, an awkward load for a small yacht with little deck space, and soon the *Lucent* looked like a Thames barge. We weighed anchor at 3.30 p.m. and moved slowly past the Angermeyer's lava-block house and out of Academy Bay. The old fridge, all the way from the Bass, bobbed along behind in the dinghy. I don't recall turning in at all that night. We sat in the cockpit, chatting quietly to Sandy and Dave. Sandy was a tough, desiccated little New Zealander, still on his way home after the war! He had called in at the Galapagos and somehow never got any further. We plodded on under power, on dead reckoning. There certainly was no feeling of exhilaration, nor for that matter of apprehension, although the enormity of our venture really came home to me during those few hours. En route it had all seemed hypothetical but now it felt real. A small boat at night, on a limitless ocean, is a powerful purveyor of mortality, especially in the small hours. June was extremely quiet, I hoped merely pensive and not regretful but it was no time to enquire.

We sighted Tower at 5.30 a.m. on 28 December 1963, and dropped anchor well out in Darwin Bay at 7.35 a.m. Whilst we breakfasted we gazed apprehensively at the breakers piling across the mouth of the small landing beach before smashing thunderously against the flanking cliffs. The distant view of our chosen island had been distinctly dull. Tower appeared long, low and unshaven, with a fuzz of sun-bleached scrub. Broken cliffs of black lava rose from the depths of the bay, which is actually the caldera of a volcano that has slipped sideways so that the far rim (the entrance) is below sea level. At the head lies a low coral beach not much more than a hundred yards long. This, barely above high water mark, was to be our home. It looked impossibly tiny from where we were – just a pocket handkerchief.

David decided that a trial run with a light load would help us to size up the dangers of a capsize. None of our boxes were watertight! We managed the first load without mishap, though the backwards drag of the retreating surge on the shelving beach sucked the coral sand from beneath our feet with such violence that we nearly fell flat on our faces. On the way back the outboard conked out. We had only one oar and Dave had prudently anchored more than quarter of a mile from the beach. I lost count of the number of trips it took, hoisting the heavy boxes over the side and down into the bobbing dinghy, then the slow row in and out again, under the equatorial sun. I'd craved sun and I'd got it, nor was I sorry. And I hugged myself every time we landed, for the beach was obviously a super camp site, surrounded by seabirds and backed by a crystal clear pool that filled and emptied with every tide – the finest private swimming pool in the world, complete with tropical fish, sea-lions, and (with June) mermaid.

In her quiet, competent way, June concocted lunch behind a chaos of crates, jerrycans, sacks and bundles. Then, just as everything seemed perfect, our hearts sank. A small Ecuadorian fishing boat with a crew of seven men 'put-putted' in under engine power and, to add insult to injury, anchored within spitting distance of the beach which we had toiled so

139

The Tower Island campsite (*above*) at the head of Darwin Bay, and the solar still that provided drinking water from sea water (*below*).

strenuously to reach. If their presence seems a small matter, it did not strike me that way. Only once in a lifetime can anybody sail up to their very first desert island. It is a unique experience, to be savoured. To have a gang of fishermen muscling in on the very first night was like a third party on a honeymoon. Later, when they had departed with their load of sun-dried fish we went for three months without sight or sound of man, boat or plane – alone in the world.

I am ashamed to admit – and maybe my tired state was to blame – that I harboured dark suspicions. Our mountain of goods lying on the beach appeared horribly vulnerable in the black of night. Every few minutes I peered furtively through the tent flap (we were sleeping in the small store tent for the time being) straining to distinguish between the barely discernible boulders and bushes and the creeping shapes of the stealthily approaching villains.

The next day we proudly erected our large, tropical canvas safari tent with supports like telegraph poles and huge cast iron pegs. No wonder the porters had staggered under the weight. It took us all day to get it up, but when we had finished we had a home beyond our dreams. Across the lagoon we looked onto nesting frigates and boobies. In front, the coral beach, opening onto the wide circle of Darwin Bay, expanded at low tide and diminished almost to vanishing point at high. Galapagos pintails visited the lagoon and night herons stalked its edges like evil gnomes. Wandering tattlers played tag with the tide, just like our familiar sanderlings, and for garden birds we had Darwin's finches, mocking birds and doves. The Bass had been climaxed.

Our island was waterless. We had known this in advance and, as I've already said, we took the makings of a solar still. Astonishingly, it worked. We dug a shallow rectangular pit in the sand, lined it with black polythene and filled it with seawater. Over this we erected a wooden frame like a greenhouse roof. We then stretched our transparent mylar over the frame, bending the bottoms inwards to form troughs. We set the whole structure on a slight slope and sank two cans into the sand at the lower end, to receive the flow of water from the troughs. The sun had contracted to do the rest. All we had to do was sit back and watch the droplets coursing down the inside of the roof, coalescing into satisfying rivulets and sliding imperceptibly into the trough.

However, the sand was riddled with small crabs which punctured the polythene base, draining out the seawater. We couldn't simply lift the roof because in order to seal it, the base had been carefully packed with sand. Instead we unpinned one end and June wriggled inside to fix it. Unfortunately I've lost the photograph of my naked wife inside a transparent cage. At best, we managed five or six pints a day – enough for survival – and an expedition with better-made stills could easily increase that ten or twenty-fold.

We had barely settled in before a small, coal-black finch with an enormous beak hopped into the tent. He was a cock large-billed groundfinch. Straightaway he perched on our fingers and accepted water from a spoon. He adored almonds, taking bits with the utmost delicacy from our lips. He was probably outside his territory for he used to steal in, and only when safely back on his cliffside cactus would he burst into his simple jingle. He came to waken us in the mornings, nipping our noses gently. Several months after we first left Tower we returned for a brief visit and who should fly down to the beach to greet us but our little black finch.

The finches, doves, and mockers were a persistent trio, particularly the doves and mockers. Galapagos ground doves are cinnamon with black markings, permanently clownish on account of the cerulean blue rings around their staring eyes. They pottered around tamely, suddenly shooting off with a dove-like clatter of wings. The mocking birds, though, were a different proposition, not merely tame but buccaneering, reducing June to literal tears of frustration. We stuffed every possible entrance hole into the tent with socks but they simply pulled them out and after a beady-eyed inspection, head cocked, they slipped in and began their thorough and destructive exploration. They were merely being their opportunistic selves, a trait essential to their survival in this hostile place.

Alas, I ignored a rare opportunity to study the social behaviour of mocking birds. They lived in groups, recognising their fellows and discriminating against neighbouring groups. Within each group was a clearly defined hierarchy or peck order. Later, on Hood Island, we discovered that they used the well-developed facial markings as recognition features.

New Year's Day, 1964; the beginning of 'the' year. It was

142

fortunate that we had a theme already in our heads, for on Tower the distractions and choices were legion. In addition to all the seabirds, in themselves far too many to encompass, there were all the landbirds, to say nothing of iguanas, crabs, sea-lions, fur-seals and innumerable insects. This still left the fishes untouched. Yes, we did need the discipline of a theme, and we had one. My gannets had told me what questions to ask and the only problem was to find time and energy enough to pursue them, in addition to filming, writing and living.

We were usually roused by the strong sun lighting the green interior of the tent which, by chance, was the same size as the Bass hut had been but even more cluttered, for much of the year's food was stored inside. A pair of swallow-tailed gulls had made their nest-scrape against one wall and their intimate vocalisations – gutteral croaks and weird clickings – ushered in each dawn. The male red-footed boobies, beginning another day's courtship before it became too hot, flew in to their bushes with rapid, tinny, accelerating chains of sound. The hoarse 'yak-yakking' of a frigate beating in towards its fully-feathered offspring, which set up a fearful caterwauling and squealing like a stuck pig, a short burst of song from our finch in the cactus. There was no sleeping through all this, even if we'd wanted to, and in any case the tyranny of the spring-balance, vernier calipers and measuring rule immediately asserted itself. The rounds had to be done before breakfast.

The loo was a sea-washed crevice – simple, clean and effective so long as you didn't fall down the crack. Peering down, and if the light was just right, the gaudy parrot fish could be seen browsing on the faeces, baring their hideous false teeth. A quick dip in the warm sea and no need for a towel – the sun dried us nicely. Breakfast was enjoyable – oatmeal with reconstituted milk, cold from the fridge, bread baked in our flat-wick oven, tinned butter and marmalade and tea. The mocking birds loved breakfast time too, and did their utmost to share it, flying up and snatching the bread from our hands. There was no house cleaning to do, other than the odd foray after giant centipedes that, in obedience to their thigmotactic, geocentric and anti-photic tendencies, had crawled into the tent and squeezed into crannies. They were the bane of my life. I shuddered simply to see them

143

trailing their chitin-armoured bodies raspingly across the floor, claws scratching and formidable jaws conspicuous on the plated head. The biggest measured eleven inches, broad to match and were venomous. Nothing I have ever seen scared me as these creatures did. Even decapitated, they ran like Japanese express trains. I think the fisherman's catch, drying in the sun, attracted them. I wouldn't have been surprised to see one making off with a booby chick.

11

Life on a Desert Island

There is a pleasure in the pathless woods,
there is a rapture on the lonely shore.
There is society where none intrudes
by the deep sea, and music in its roar
Byron

We were still piqued by the mere presence of *San Marco* and
her harmless crew. They were virtually in our backyard,
interested in our every movement, constantly coming ashore
right in front of our camp to collect driftwood for their
brazier or to stretch their legs. It was the more galling because they had more right to be there than we had! They
were certainly hardy and industrious. Seven men lived,
worked and slept on a battered little boat scarcely more than
twenty feet overall. A wooden cradle projecting over the
transom held a charcoal-burner brazier for cooking. Amid-
ships stood a fifty-gallon drum of seawater in which they
washed their filleted catch before laying it out on the coral
beach to dry in the sun. At grey five o'clock every morning,
they left their mooring and puttered slowly out of Darwin
Bay with their booby escort, to the western side of the island.
Occasionally, when we visited the masked boobies on that
side, we saw them tossing up and down in the chop, fishing
endlessly. They returned in mid-afternoon to gut, salt and
dry their catch – another long and toilsome task. Then a
frugal meal and nothing to do but retire below decks in a
squalor which I could merely imagine, for I never saw it,
before the next grey dawn. They were a swarthy, ragged and
villainous-looking crew, but as pleasant and courteous as
could be, never coming for treatment for cuts, boils or tooth-
ache without a thank-you fish. They fished from Darwin Bay

145

for about six weeks and then, hoisting a filthy rag of a sail, set off back to Wreck Bay, finding their way without a compass, let alone a sextant, and relying (perhaps wisely!) on the efficacy of matchstick crosses tossed overboard, rather than mechanics, in the event of a break-down. I would have been scared stiff to go with them.

We had been on Tower scarcely a week before another visitor scattered our composure. This time it was a big black yacht of sinister appearance with a black-bearded crew to match. Francis Mazière and his Tahitian wife (unbearded) had already been a year en route to Easter Island, filming as they went. His technique amazed me. He marched up to a frigatebird which, with customary Galapagos tameness, sat stolidly on its nest and when the bird's beak barred further progress, he pressed the button of his ciné camera and the job was done. If the lighting was acceptable, for it was blindingly bright, and the shakes hadn't been too bad, he should have managed a splendid sequence in glorious technicolour of a frigatebird doing absolutely nothing. I have already mentioned that Niko Tinbergen's passion for photography had rubbed off on me (oddly, David Lack never took photographs). On Tower I developed my black and white stills in a daylight developing tank, though many of my negatives ended up both dirty and marked. It was difficult to control the temperature and we were desperately short of clean water to wash the film. All the rinsings, already green with algae, were used for June's hair wash. Thanks to the generosity of the Frank M. Chapman Foundation I had a 16 mm Bolex but I hadn't enough film for general filming, so to my lasting regret, I was unable to put together a popular Galapagos film though my efforts did stimulate Aubrey Buxton to send out his Anglia 'Survival' team. The result was not only a prize-winning film by Alan Roots but a Royal première and a magnificent donation to the Galapagos Foundation which enabled them to buy a new research vessel to replace the ancient *Beagle* of beloved memory.

Over a meal on board, the Mazières talked about their friend Dr Bombard, the famous French doctor who crossed most of the Atlantic alone in a rubber dinghy without food or water other than that yielded by the ocean. This amazingly courageous exploit was carried out, against much official hostility, to demonstrate that ship-wrecked mariners could

146

Lava gulls.

survive on a life-raft in tropical seas without food or water on board. Nobody expected Bombard to come out alive. He should have made a modest fortune but the terms of his book contract were apparently such that he received a pittance, and at the time of the Mazière's visit he was reluctantly teaching history in a small French town. I bet he gazed dreamily out of the window more often than his pupils did. We took the Mazières to the north-west corner of Tower but it was a hard grind, for they chose the hottest time of day, when the interior of Tower was bakingly hot and sulphurous. The brittle, razor sharp lava sheets broke and cockled with every step, hell on the feet and agony on the ankles. But the silver Palo Santo trees were in tender leaf and bore a few lovely yellow flowers. Emerging into open ground we discovered a large colony of masked boobies, the whitest white seabird in the world. Sooty lava-gulls with flame-coloured gapes dive-bombed us and must have had nests nearby, but the expanses of rust-streaked black lava effectively concealed them. In a nearby lagoon a forlorn young sea-lion lay with a fore-flipper torn off at the shoulder, no doubt by a shark. The offal thrown overboard from the *San Marco* attracted

147

dozens of sharks and today the fishermen hooked a black-tip and dragged it ashore. When they had dismembered it and the coral sand ran red, the lava gulls, much like a small dark herring-gull, scavenged for bits but the swallow-tailed gulls completely ignored it. They feed mostly at night, on squid.

During our very first month we were invaded by twelve American biologists, part of the celebrations attending the official opening of the Charles Darwin Research Station on Santa Cruz. 1964 was also International Biological Programme year and the visitors were supposed to come up with fertile ideas for future research in the Galapagos. The group should have included Niko Tinbergen but alas, as the Ecuadorian naval boat, suffering from rudder trouble, wove an erratic course into Darwin Bay the binoculars showed that Niko was not on board. He was one of the very few visitors that we really wanted, with his ideas, enthusiasm, fun and insight into animal behaviour. One of the group had been a doctor on Samoa and was investigating the anaesthetic properties of an extract of sea-anemones. Samoans have an effective and painless method of suicide. They eat a particular species of anemone which first anaesthetises and then kills. At that particular stage I felt a passing personal interest in this discovery. My tropical sores were still unhealed, I had a crop of mouth ulcers, mild dysentery and seemed to be collecting more than my share of tension-headaches. Also, my old Exakta was up to its infuriating tricks again, refusing to re-wind and tearing film in the process. It had been for repairs three times before the trip to obviate just such a disaster.

Fortunately, the spate of visitors soon dried up. This sounds churlish but one of the delights of living on a desert island is this very freedom from intrusion, whether from visitors, radio, telephone or any other distractions. It is an artificial and unsustainable freedom because it depends on outside help, but as a temporary retreat it offers great rewards to those of a compatible frame of mind. It is a curious feeling, this detachment, which cannot be matched by a brief holiday in remote surroundings. As the weeks stretched into months, without sight or sound of man or his artefacts, we felt more and more self-contained, although in fact we could hardly have survived a month without the food which we had brought. For me, the whole venture would have been ruined

148

had it involved other people. If we had been part of a normal expedition, a group, I would have found it a real strain, whereas with just the two of us, the days and weeks slipped peacefully by with never a cross word. But equally, I would not have even contemplated such a venture on my own. Our particular circumstances were ideal and it was my good fortune in this respect that made the whole thing a success. For many, the future must stretch drearily ahead with its procession of utterly predictable and boring days. For us, it was all to play for, full of potential. In the evenings, with the old Tilley lamp shedding a pool of yellow light against the tropical darkness, we spent many a happy hour day-dreaming amidst the comfortless clutter of our crowded tent. We were virtually penniless, with no home, or furniture, car or clothes, so we had plenty of scope for material day-dreaming. I wanted a Land Rover, and much later I got one. It took us on many memorable trips, through the loose sand of Wadi Rhum, the dramatic setting of 'Lawrence of Arabia', over flash-flooded desert with the water half-way up the doors, across miles of black basalt shimmering in such heat that even the desert-adapted larks looked on the point of expiry. It survived sea-crossings, frontiers, impoundments and stoning, carried jackals, jerboas, pi-dogs and a wolf to say nothing of two members of the British Royal Family. At times we lived in it, short wheel-base though it was. It ended its life on an Aberdeenshire farm, God rest it.

One of the simple pleasures of life on a tropical desert island is to throw off one's clothes. To swim in a warm sea, dry in a hot sun and go peaceably about one's camp chores in this natural state is a delightful freedom that only those who have tried it can appreciate. It is sad, but given our unlovely society perfectly understandable, that we view nakedness with suspicion and embarrassment. But it is a pity. Interestingly enough, it seems to be the Europeans, and especially the Germans, who seem sun-starved enough to grab every chance to soak it up in this way. The fishermen of the *San Marco* would never dream of it. They get too much sun. The Italians and Spaniards too, are much less likely to go naked than the Scandinavians, so religion probably has much to do with it, too.

After a couple of months we felt much more relaxed. A work routine had been established and the notebooks were filling up nicely. Between the stained covers and buried amongst the scribble there was a scientific story, maybe even one or two small nuggets. The reader may wonder what is to prevent a fieldworker from concocting his data and spending his time lazing around. Who is to check up on what you did, as against what you say you did? Almost always, though, the fieldworker wants to know the *real* answer and I doubt if one in a thousand even thinks of the other point, which is that they may be found out. If you publish facts and figures, another worker may well replicate them. If they don't match, it may seem interesting enough to probe further and then anything could happen. Anyway, we had settled well into our desert island life and hard work.

Nowadays we cooked supper on an open fire under the velvet sky. The sea surged rhythmically against the coral beach, soothing and only half-heard, great Saturnid moths, like bats, stroked the warm air and the gigantic centipedes foraged like the predators they are. On the fire we baked new bread in a biscuit tin packed with coral fragments to spread the heat. That homely smell seemed faintly incongruous there. Once, we boiled lobsters to go with the fresh bread and baked a chocolate cake. Then we danced on the sand to fuzzy music from Radio Belize. I write all this, factually correct in every detail, but am quite unable to recapture even a glimmer of the feelings that accompanied it. This, to me, is a minor tragedy – that so much of our pleasant past can not be totally re-lived in memory. Maybe some fortunate souls can do so, but I can't.

In early February a few rain-showers transformed the island, throwing a delicate green veil over the bleached skeletons of the driftwood-dead trees. For a brief fortnight the black and rich brown of the lava was softened by this evanescent foliage and the air deliciously scented. Then the leaves died and Tower became a sun-baked desert again. It became a real effort to cross to the far side of the island, carrying ciné camera, telephoto lens, heavy wooden tripod, Exakta, film, ringing and weighing equipment and notebooks. By the time we had shot 400 metres of film, caught, ringed, weighed and measured fifteen adult boobies and twice as many chicks and then humped the gear back again we were more than ready to jump into the lagoon and soak, and drink five or six pints of liquid.

150

The rain set the Darwin's finches singing and courting. The doves were poor singers but they could court. The male flicked his wings so that the tips were barely raised and then trembled them, at the same time fluffing himself out enormously. The female became similarly rotund and the pair then shot into a tall, upright posture, gripped beaks and with circular blue eye rims imparting a foolish, fixed stare, twisted each other's head round and round as though trying to unscrew it. Bouts of preening and 'false feeding' intervened during which they pecked at the ground without picking anything up. Displacement activities such as these are common in courtship or any other 'tense' situation and are indeed essential if behaviour is to run its normal course. They are 'spacers', giving animals a chance to reorganise internally. We frequently do the same without recognising it.

By late February the island was vibrantly alive. The rain had stirred up the land birds, the frigates and boobies were displaying and courting, night-herons building and even the small, coal-black iguanas had begun to dig burrows and lay eggs. On one tiny beach, about ten by thirty yards, well over a thousand had congregated as motionless as lumps of coal. But whereas in temperate climates such things proceed without hindrance, in the tropics there is no such guarantee. In a quarter of a century, I have never known a single year in which the Bass gannet's breeding cycle was disrupted or even slightly disturbed by natural disasters. On Tower, the day was not far distant when breeding activity was to be stopped in its tracks because food ran out, only to pick up, as suddenly, because food reappeared. Around the Bass, the seasonal movements of mackerel, herring and sand-eels are predictable and the gannets have been able to fit their breeding activities into this pattern. But when food appears and disappears unpredictably such dovetailing cannot evolve and famines are inevitable. God may look after sparrows but frigates must look after themselves.

By early April the thick-billed ground finches were feeding fledged young. Each male appeared to tend a single juvenile which followed him around, begging persistently. The adult regurgitated food piece by piece, with a quick head-jerk. The conspicuous cream-coloured lower mandible of the juvenile perhaps acted as a marker in the gloomy thickets where they fed, much as the young swallow-tailed gulls's pale beak spot does in the shady places in which they often shelter.

The weeks slipped past and on 5 March our 'relief' arrived on the *Lucent*, with fifty gallons of paraffin, forty gallons of water and fresh fruit and vegetables from the generous Carl Angermeyer. After drinking our next six-months' supply of beer, everybody departed in good fettle and left us to our peace. Stimulated by the 'relief' we set-to and cleaned up the camp. Three months'-worth of empty tins were pierced and taken out to sink in the abyss of Darwin's Bay. All the juvenile boobies trooped out with us.

My birthday falls on 14 March and on that day we took a picnic in the boat. First we rowed along the west side of the bay but it was rough and we couldn't land. Several sharks and an enormous Manta ray swam directly beneath us. Overhead, a greater frigate was attacking a fledgling Audubon's shearwater, repeatedly picking it up and dropping it into the water. Out in the mouth of the bay – which we avoided at all costs because of the current – a herd of dolphins fed in water-churning rushes. After rowing across inside the mouth we found a negotiable crack in the cliffs. It emerged onto a great plain of volcanic ash and fragments of clinker upon which stood a few large boulders and stunted trees. Further round the North Horn of the bay we discovered a vast field of weathered lava, utterly bare and desolate, a broken, wafer-thin crust with piles of broken scoriae piled haphazardly like millions of smashed plates. Above this shattered crust flitted thousands of storm petrels in mazy, erratic flight. The air was thick with them but the only sounds were slight scratchings and the muffled rasping of wings as, every few seconds, a petrel landed and slipped into a crack, or emerged from the crust. Frequently, they pattered over the lava with dangling feet, as though over the surface of the sea, before slipping quickly into a crevice. Later, with David Snow, we collected some of the many wings which had been parted from their owners by the fierce little Galapagos short-eared owl, whose 'parlours', containing mounds of feathers, we found in larger recesses. We discovered that there were two species, the Madeiran storm petrel and the Galapagos storm petrel. The meaning of this mass flighting behaviour is still a mystery although we do know that on some of our own Scottish islands, where storm petrels fly around at night in large numbers, most of the birds are non-breeders and there is much visiting between islands. But the benefit to the petrels for this risk and the expenditure of energy remains unknown.

Three yachts dropped anchor in Darwin Bay whilst we were there. The Mazières I have mentioned. In 1964 the *Observer* single-handed Atlantic races had barely started and certainly people had not yet made a habit of sailing across the Atlantic in bath-tubs. *Popeyduck* (Cornish for *Puffin*) with a length overall of 20'8", beam 6'3" and a registered tonnage 2.46, was at that time, I believe, the smallest boat ever to attempt a single-handed voyage around the world. She had been built in the garage of an Oxford suburb by her owner, Bill Proctor, to a design by Laurent Giles. I have realised, since, that it is not always the tough, extroverted, machismo type that goes in for the really big lone adventure, but often the sensitive, almost visionary person. Bernard Moitessier, for instance, who circled the world once, in *Joshua*, and then carried on for a second time round to 'find himself'. But at the time I simply marvelled that Proctor, a man aged 51, a refined intellectual in indifferent health, thin and stooped, could have built such a boat in his garage and then had the courage to set off round the world in it, alone. Could this bespectacled ex-civil servant really cross the world's great oceans, meeting gales in the blackness of night, coping with mountainous seas, damage to the boat, fatigue and sleeplessness, and nobody even to make a cup of tea? Think of it when you lie in your comfortable bed on an appalling winter's night, listening to the wind-driven rain pelting the windows. For his part, he was amazed to spot a tent on an uninhabited island in the Galapagos, and a little wary of investigating!

Once we had made contact he spent every available hour ashore and there were few unavailable hours. June's cooking was a great attraction but he threw himself energetically, if a trifle clumsily, into booby-catching, suffering a few painful bites. The life of seabirds was the only subject about which I knew more than he did, for on everything from bee-keeping to boat-building, via bread-making, the law, politics, art, literature, religion and beer, he was irritatingly erudite. Contrary to what some modern educationists seem to think, there is simply no substitute for a well-stocked mind – not that I can claim one. Tragically, Bill met his lonely end when *Popeyduck* was smashed to matchwood on a coral reef in the

153

south-west Pacific. Probably he misjudged his landfall and hit it in the dark. I often think of those final minutes. It seems a stiff price to pay for courage and enterprise, but one he was prepared to pay.

Perhaps less notable, but equally laudable, were the owners of *Kismet*, a Bermudan-rigged sloop of around thirty-five to forty feet. Two American couples had pooled eight years of hard saving to buy the boat and finance a world cruise. I can recall only two occasions in my life on which a quiet inner voice told me that I was definitely not in control of my motor-coordination. One was after the end of an Ethological Conference at Starnberg where an inexhaustible supply of German sausage and beer slipped insidiously aboard whilst we listened to a Bavarian folk-band in the courtyard of a medieval castle. The other was on the *Kismet*. The owners invited us aboard and then slyly tipped the fuel from their alcohol-stove into the orange juice. The effect was unfortunate. Sitting in the cockpit, watching the tropicbirds and eating popcorn, the orange juice seemed innocuous enough but I rowed ashore like a whirligig beetle and feeling about as intelligent. It seems so much more imaginative to spend money as they were doing rather than on cars, gadgets and booze, but it needs competence and a spirit of adventure, and modern society kills both. Like J. B. Priestley, I am not in love with modern society.

One of the four was a doctor, and she was interested in our health. If we were to repeat the venture I'm sure we could greatly improve on this. We both were lethargic and, as I've mentioned, I suffered from boils and mouth-ulcers which could have been avoided by dietary supplements. Modern diet, despite the assurances of Miriam Stoppard, is woefully deficient in many essentials. In addition, as we are increasingly aware, it is dangerously contaminated with additives, with which our tinned food was laden. At the time I felt aggrieved that we were not bursting with life and energy, since we enjoyed such a marvellous life, but I now understand why. If I were doing it again, I would certainly take big daily doses of vitamin C and small amounts of trace elements such as selenium and zinc. As our energy waned, our artifice bloomed. We introduced 'efficiency days' on which we were supposed, each in turn, to overcome the back-log of unfulfilled tasks. 'It's your efficiency day' was the smug cry of the

154

Beagle II beached for bottom-cleaning.

off-duty partner. For me, it meant developing films, mending the solar still, trimming the fridge wick, writing an article or whatever, in addition to the daily bird-checks. For June it might be baking, patching and sewing, washing her hair or tearing my manuscript to pieces. Pre-menstrual tension sometimes upset the efficiency routine by introducing temperamental days, on one of which I recorded June snoozing on my bed after swallowing one askit powder, a mint, eight prunes, two pieces of marmalade and bread and two cups of coffee. Things did tend to get depressingly behindhand. Often we returned tired and hungry from the late-afternoon round of weighings and observations, just as darkness descended, to find the tent littered with half-finished jobs, all our crocks still to wash and the supper to cook. It required a real effort to deal with it all. There was no light, since paraffin was low, very little water, insufficient utensils, dreary food and a mess everywhere. At such times, Tower did seem grey.

But on 1 June, our depleted larder was enlivened by the addition, courtesy of *Beagle II* of the Charles Darwin Research Station, of three pounds of fresh turtle meat, from which June made a superb meat and potato pie. We left Tower in the *Beagle* at 5.30 p.m. on Friday, 10 July 1964, after a stay of 194 days. During the night the current threw us badly off-course and by morning we were heading back towards Marchena, only twenty-five miles away. I believe Julian Fitter had lashed the helm and snoozed. Carl Angermeyer, the skipper, was not amused. The Angermeyer story has often been told, but bears repeating. Five brothers fled Hitler's Germany in the late thirties. Hans died of kidney disease in Guayaquil and Henry returned to his home. Carl, Fritz and Gus carried on and settled in the Galapagos where they remain to this day, the most famous of its inhabitants. House-building, boat-building, farming, fishing, painting, writing, guiding visitors, skippering charter vessels – it's all grist to the Angermeyer mill. Hospitable to a fault, they must have entertained more visitors than there are spines on a cactus. The day following our return to Academy Bay Carl invited us for a meal. Inside the lava-block house the furniture gleamed, tableware was elegant and the squalor of their first years on the island must have seemed a distant memory. But after a few days of socialising we were ready for the next leg, a four-month stint on Hood Island in the south of the Archipelago. Before that, though, I have some seabird stories to tell.

12

The Boobies and Frigates of Tower Island

Dampier had a true scientist's curiosity and an impressive determination to observe and record facts as a preliminary to deduction. The new fashion for scientific enquiry had been greatly stimulated by Charles II's foundation of the Royal Society and his real personal interest in many branches of science. Dampier was almost a contemporary of Isaac Newton and he dedicated his 'New Voyage around the World' to the President of the Royal Society when it was published in 1697. The success of this book, which gave the world the first printed account of the Galapagos Islands, brought Dampier a considerable reputation and a great deal more money than his efforts at buccaneering had done.

Hickman, *The Enchanted Islands*

When we sailed into Darwin Bay, scores of red-footed boobies winnowed above the *Lucent*, as their ancestors had done above Darwin's *Beagle*. But the effect was puny compared with the multitudes of gannets on the Bass. If anybody had told us that there were maybe ten times as many red-feet on Tower as gannets on the Bass we would have ridiculed the idea. Yet when we estimated their numbers, it seemed to be so. The theory was to select a few squares with sides 100 yards long and count the nests or nest-remnants in them, take the average and then multiply-up for the whole island, leaving the non-vegetated parts out of the calculation. This procedure is full of pitfalls. Even counting nests in 10,000 square yards of spiny shrub in rough terrain, with deep fissures and the like, is not simple. We actually cordoned-off our squares with string but since we didn't use a compass it sometimes took an hour or more even to find the starting

A pair of Galapagos red-footed boobies:
brown morph and white morph.

point in order to join-up. We even lost each other in the
process. When we had completed the job and multiplied-up
we ended with the amazing total of 140,000 pairs, making
Tower the largest colony of red-footed boobies in the world.
This must surely have erred on the high side, probably
because it included redundant nests.

Tower, indeed, embarrassed us with riches. The two
boobies, masked and red-footed, claimed most attention but
it was impossible to ignore the frigatebirds and the elegant
swallow-tailed gulls, both of which nested outside the tent.
To the shearwaters, storm-petrels and tropicbirds we did,
however, turn a blind eye. We had quite enough on our plate
documenting the breeding behaviour, especially the displays, of the boobies and frigates, recording the timing of

laying, the pattern of attendance at the site, the growth of the chicks, the number of deaths and their causes and the details of the gradual acquisition of independence by the young. Really, we wanted much more, but only a longer study with recognisable individuals could have told us about adult death-rates, fidelity to site and mate, how often they bred and much else. All this, and the mundane details of nesting habitat, density, egg characteristics and so on were perfectly straightforward. It was simply a matter of time and effort.

Food is almost always the big unknown in seabird studies. It underpins the entire breeding strategy but is always difficult or impossible to observe directly. The important aspects are whether food is plentiful or scarce, near or far from the breeding place, seasonal and predictable or erratic. To all these facets seabirds have had to evolve complex adaptations. As our daily checks on the attendance of adults and the growth of chicks mounted we began to see, although dimly, an underlying pattern. Soon we could recognise the spectre of approaching famine before the sad tally of dead and dying chicks and deserted nests. The first hint that all was not well came when those adults which were incubating or guarding chicks had to remain on duty longer than usual. This meant that their mates were experiencing difficulty in finding food at sea. In the cases of the boobies, five or six days was about as much as they could stand, but frigates endured twice or almost three times as long – so long that I developed a positive aversion to seeing the same drooping bird day after day. During these punishing spells they lost a fifth of their weight. As for the chicks, they either gained weight painfully slowly or not at all. It was not unusual for a red-footed booby chick in a famine to weigh less than a third of its normal weight. Many simply wasted away and died and the colony assumed a characteristically derelict air. But the sudden upturn in fortunes was just as dramatic. Chicks, pulled back from the brink of starvation, began to put on weight and it was business as usual again. Clearly, important changes were occurring at sea. It seemed obvious, and still does, that the birds themselves could not possibly have been responsible for the sudden shortage of food, if only because the return to normality happened so suddenly and when there were just as many birds as before. I imagined that shifts in ocean currents, or in the position of convergences and upwellings,

Fratricide in the masked booby (*above*). The brood of two is reduced to one by the first-hatched evicting or killing the second-hatched. This avoids the possibility of the parents rearing two weak chicks, whilst retaining an insurance policy against the early loss of the first (the second hatches 4-5 days later). (*below*) The survivor.

which are the localities at which nutrients, and therefore phyto- and zoo-plankton, and thus fishes and squids, are abundant, or changes in temperature, had been responsible for the famine. The devastating effect of sea temperature on marine life is evident when warm water from the Panamanian region invades the area normally cooled by the Humboldt current. The greatest sufferers are the millions of Peruvian seabirds whose breeding islands are normally washed by this cold, rich water, but the Galapagos Islands are by no means immune. The Humboldt does skirt the archipelago and has a marked effect on marine life. In bad years, not only Galapagos seabirds but also marine iguanas, seaweed and marine invertebrates suffer severely.

Ecological work can be a real chore but behaviour studies never seem to be, I suppose because the animals are always doing something. Curiously enough, it is much harder to compile a detailed inventory of an animal's behaviour than one might imagine. If five people set out independently to weigh and measure birds their results would agree closely. But five accounts of the same behaviour would probably vary widely. This is partly because units of behaviour can be difficult to demarcate and also because it is so easy to be subjective. I like to think that, from my descriptions, anybody could recognise the various displays performed by gannets and boobies, but I know that another person's interpretation might have differed from mine. The absolute essential is precision, which means unambiguous descriptions of the form, amplitude, frequency, context and sequence of clearly defined acts. We need this kind of description for 'normal' human behaviour, but what we often get are highly subjective and imprecise accounts of abnormal behaviour and quite incoherent studies of social behaviour conducted by people who have insufficient ethological background.

Seabird colonies are, on first impressions, noisy and chaotic and the many happenings have to be sorted out before they can make any sense to us. Take, for example, a kittiwake colony. The kittiwake is not going to fly into its site, neatly fold its wings, pause for a decent second or two, launch into a discrete 'piece' of behaviour, pause whilst you record that, and so on. Its actions will occur in rapid sequence, often interrupted by something its mate or a neighbour does. The break-points have to be determined before one can talk about

161

'Choking' in the kittiwake (*above*) displays the bright gape
and is used only on the site. (*below*) Post-landing and
'upright' posture in the kittiwake.

behaviour A preceding B or eliciting response C, otherwise
one doesn't know what behaviour A or B *is*.

If one purports to be compiling an ethogram, as a list of
behaviour is called, it is essential to distinguish between
one-off acts, which are not fixed in form, and repetitive,

stereotyped behaviour which has a 'shape' and is easily recognisable every time it occurs. It is this kind which is used when animals communicate with each other. It has a 'message' which is readily received and understood, which means that it works most effectively when it is clear-cut and dependable in form, as displays are.

Then there is the problem of demarcation. Where do you place the boundary line of an act? Is the unit of behaviour to which you want to give a name in fact two (or more) discrete acts or simply one act with two (or more) components? Usually, prolonged observation will show either that the 'acts' are tied together and are therefore components, or that they usually are not. But in some cases acts are certainly not strictly tied together but nevertheless *tend* to occur in predictable sequences. So long as the actual situation is faithfully described, the observer has done his job and can interpret it as he thinks fit, leaving his peers to make of it what they will. I have no interest at all in subjective and unsupportable interpretations of animal behaviour but equally I consider that an ethologist has done only half the job if he or she provides no informed guidance about the probable function of the behaviour. We want to know why an animal behaves thus, and what the behaviour achieves. These points are old hat, long since expounded by European ethologists, if less so by Americans. Many will point out that ethology has moved far ahead of plain observation but we are now aware that a very great deal can be extracted from, indeed depends upon, simple but detailed and accurate fieldwork.

The study of animals in their natural habitats is far from redundant. The significant effects upon survival and reproduction of different ways of behaving – even apparently trivial ones – means that we cannot begin to understand *why* animals behave in the way they do without knowing a great deal about their behaviour. And the more species we know about, the sounder will be eventual generalisations. Usually, really basic generalisations about behaviour ultimately suggest things about human behaviour, as the furore which attended the sudden prominence of sociobiological theories so vividly demonstrated. This is because the more basic the truth, the wider its application. Soon, if they are not doing so already, sociologists will have to pay more attention to biology.

163

The Galapagos boobies were naturally of enormous interest to us because of their close relationship to gannets. This is not the place for a dissertation on comparative behaviour but I will try to recapture the process, not without its growing pains, of assimilating the behaviour of the red-footed booby and fitting it into place within the framework erected for the gannet. So off I went to sit in the sun with stopwatch, binoculars and note-book. It is a mistake to begin detailed note-taking immediately. There should be a tuning-in period during which you simply watch and listen. Fortunately the red-feet were nesting in low shrubs behind the tent, so it was delightfully simple to watch them. Later, when I saw red-footed boobies high in the jungle trees of Christmas Island and in the green depths of the mangrove thickets of Aldabra I realised just how lucky my choice had been. At first, things seemed rather dull compared with the frenzied activity of the Bass gannetry in full cry. There were a few ever-watchful frigates high in the sky and, just beyond the shrubs, jammed in the fork of a tree, the dried carcase of a red-foot that had hurled itself pell-mell to apparent safety with a piratical frigate hard on its tail. There were a few red-foot nests with chicks, but these were not what I wanted. I had to find some pairs just setting up home.

You will be luckier or more observant than I was if you see a red-footed booby house-hunting. They do not hop around in the branches trying out various spots, in the way blue-tits so obviously search for suitable holes. But they do betray their intentions, or more precisely, the stage they have reached in the breeding cycle. Ordinarily, a red-foot simply flies in, brakes and drops onto a twig or branch but a bird in the throes of establishing its site calls loudly and rapidly as it flies in and then, still calling like a rattly old tin can, it swings its head forwards and downwards from side to side, often biting the twigs. This was interesting because it was not all that different from a territorial gannet's landing and post-landing behaviour.

Once the red-foot's territorial behaviour had been slotted into place its pair-formation claimed attention. Unlike the gannet's, it was perfectly obvious. The male on his site eyed the female perched nearby, or perhaps flying past and raising his head vertically upwards uttered a drawn-out, grating call. At the same time he raised the tips of his closed wings.

164

This was doubly interesting because it closely resembled a display performed by the gannet but in an entirely different context (a fuller account, embracing the whole family, is in chapter 15). The red-foot and gannet each had the 'same' display but whereas in the red-foot it was a sexually-motivated, 'advertising' display, with the message 'approach, I am a receptive male', in the gannet it was our old friend the skypointing display performed by partners that were already bonded, with the message 'I am about to leave my site'. Naturally, it is not enough simply to *say* that the displays are in fact the same, meaning that they both derive from the same source and are homologous in the same sense that a bat's wing and a bird's wing both derive from the forelimb of an ancestral vertebrate and are thus homologous. In the latter case, one can actually see the basic similarity in structure – the bones are there. But so, too, one can see the 'bones' of the display. In both the red-foot and the gannet the display involves lengthening the neck and lifting the head till the bill points upwards; raising the wing-tips by a swivelling movement and uttering a special call. It could of course be argued that this resemblance is just coincidence, unlikely though it may seem. But this becomes altogether too improbable when (as we later found) all the other boobies (except Abbott's) have it and all of them use it as a sexual advertising display, just like the red-foot. Obviously, the boobies have each retained this particular behaviour even though, physically and in other ways, they have split into several different species. They have each *modified* it in various small ways, but just as the structure of the forelimb remains discernible in its different forms, so does the display.

Soon after our arrival on Tower we mounted a successful onslaught against the masked boobies. In one day we caught thirty-four adults, including seven pairs, among them some that did not even possess an egg or chick to chain them to their sites and which were especially valuable because they were just establishing their territories. These powerful boobies are easy to catch and if their escape route was blocked many simply allowed us to approach and gently take hold of their formidable beaks. By dint of some athleticism one or two were caught in flight, just after take-off. Often they were so little put-out that we were able to replace them, resplendent in their coloured chicken-rings, back on their sites. It is

165

easy to see how colonies have been devastated by sailors and natives – thousands could be, and no doubt have been, clubbed in a day. In the hand they are lightweights compared with gannets – less than half as heavy although nearly as big. This difference tells a tale of adaptations to the tropics against the solid gannet's fittedness to the bitter cold and the tempestuous North Atlantic. Also, as mentioned previously, gannets need to penetrate deeply after a dive, to hold large and muscular prey and to carry reserves enough to last out the often prolonged stormy periods when it is impossible to fish.

So, with this sort of work to be done on several species, time never dragged. The great frigatebirds were a prime diversion from boobies. Like the red-feet, with whom they often nest (a red-foot has even been known to take over and rear a frigate chick), they built in the low shrubs and were so tame that it was easy to remove the egg from beneath a sitting bird and replace it without causing any disturbance. When we arrived in late December there were no nesting birds but the cryptocarpus was dotted with rusty-headed juveniles, hunched like witches on their perches and periodically squealing in a grating pig's voice whenever a parent showed signs of landing nearby – hoping they would deliver a semi-digested squid or flying fish. These huge juveniles could fly perfectly but they still depended on their parents for food. Even when we left, in July, they were still being fed although many had died of starvation in the interim. It took some time before we became aware that we were puzzled. At first there seemed nothing unusual about juvenile frigates being fed by adults but then males with enormous scarlet throat balloons began displaying in the very areas in which the juveniles were sitting. Accustomed as we were to seabirds that nested once a year, we thought, perfectly stupidly, that perhaps the male parents were cutting the apron strings whilst embarking straight away on a new cycle. A moment's reflection revealed the absurdity of such an idea. How on earth could a frigate go straight from the protracted and demanding task of feeding a youngster to the even more strenuous one of displaying, mating and building. Where did moult and recuperation fit in? Again, what would be the point of displaying if you and your female were still jointly feeding the juvenile? The display was self-evidently directed towards females as a preliminary to pair-formation, but the pair were still together! No, the displaying males could not be involved in feeding juveniles, nor could they be males that had just ceased to do so. They must be a new lot. But this meant that, if the old lot had started *their* cycle at the equivalent time last year and were not beginning again this year, they must be breeding less often than once a year. It may sound glaringly obvious but sometimes the penny drops slowly. Of course we really needed marked individuals but, whilst catching them was easy, marking them was not. Their legs were so short that the tarsi were practically invisible, which meant that colour-rings, so effective on the masked boobies, were out. Nowadays some workers ring frigates above the tarsal joint. Wing-tags were too dangerous on such an aerial bird and the black plumage made dyeing difficult. We had no white paint with which to spot-mark their heads. In the end we resorted, ineffectually, to sticking-plaster on the beak but this soon fell off.

New problems then became apparent. These displaying males seemed somewhat footloose. A group of them would spring up and then, after a few days, some would disappear and a new group would arise elsewhere. Now, suppose our rusty-headed juveniles' father finally cut loose from his demanding offspring, went off to sea and spent some time moulting and resting. And suppose he returned with next year's batch of spruce, scarlet-throated males intent on display. If, as we had seen, he was not going to stick to his old site but was instead quite willing to join a display group and perhaps move around (as in fact he did), how could his old mate find him? Clearly, she couldn't. One might suggest that she flew around looking for him among the suitors on offer but this, besides seeming faintly ridiculous, clearly didn't happen. It takes two to tango, and the males certainly were not rejecting *any* of the females who came down. A male who ardently courted two or three females in quick succession couldn't have been restricting his attentions to last year's mate. So it appeared that the system of annual breeding and fidelity to site and mate had broken down under the pressures of the tropical environment, or perhaps it would be better to say that a more appropriate system had evolved.

The crux of the matter was the long drawn-out dependency of the young. If they couldn't achieve competent independence in a year, then the parents couldn't breed once a year and the remainder of our observations followed. But it is perhaps artificial to attempt to isolate one factor as being responsible for a whole web of adaptations. Other seabirds, for instance the masked booby, on the same island, eating roughly the same food and of comparable weight, manage to breed annually. The young frigate, for whatever reason, needed longer than the young booby to become a competent feeder. But it is highly subjective to say that a frigate's feeding technique is more difficult to acquire than a booby's. It would be nice if we had some way of testing the idea, but we haven't.

These thoughts, however, arose later. At the time, the frigates simply continued to spring surprises. The courtship itself was not a surprise, it was merely extraordinary. The frigate's scarlet throat sac, fully inflated with air, was tightly stretched and getting on for the size of half a football. Down each side of its veined surface ran two or three rows of small black feathers. Half deflated, it became baggy and wrinkled and in flight flopped around in a faintly obscene fashion like a bull's testicles. Frigates' wings span about eight feet, the upper surfaces an iridescent black, shot with green lights, the under surfaces silvery. These huge reflectors, with the scarlet sac couched between, were swivelled so that they faced the sky and trembled violently. As though all this was not enough, they uttered a high falsetto warble. A group of displaying males, five, ten or even twenty strong, congregated densely on the green shrubs and the white-fronted females soared overhead, eliciting perfect frenzies of display. Eventually, one circled lower and alighted delicately near to one of her suitors. There she stood, rather vacantly, sometimes literally embraced by his wings. If she remained, there followed periods of mutual head waggling interspersed with quieter interludes. At this stage there was of course no nest whatsoever, since the male's habit of pulling up stakes and moving elsewhere meant that nest building could start only after the pair had formed. This, in turn, meant that there was no point in the male frigate having a well-developed territorial display since, until he had acquired a mate, he effectively had no territory. Weaned on the male gannet's frenetic territoriality, this came as no small surprise, until the chain of circumstances explained it. Once the pair had formed and mating occurred, nest-building quickly began. The male may have been hanging around, displaying in the baking hot

168

sun for days on end and that costs energy. It behoved him to get on with the next stage. It wasn't easy to find enough twigs to build a nest, even though it is the merest platform, often as bad as a woodpigeon's. If he could, he stole them from the industrious red-footed boobies, who in turn plucked them from the trees and bushes. Breathtaking piratical aerobatics were performed for the sake of a single miserable twig.

The next puzzle came soon after the single large white egg had been laid. Mysteriously, eggs began to disappear. At one stage the situation became farcical, with eggs on the ground all over the colony. This was not a result of disturbance by us – a factor which I would be the first to suspect in anybody else's study – because the birds were so tame that we could handle the egg without putting them off the nest. Eventually, it became clear that there were spare males hanging around, and although we only rarely actually saw one interfere with a nesting attempt, we put most of the egg-loss (and later the loss of chicks) down to interference by male frigates. Once I saw an intruding male frigate maul a chick on its nest whilst the owning male perched nearby and did nothing about it. Often we found small chicks lying dead beneath their nests with peck wounds on the back. Non-breeding but apparently mature males are present in colonies of many species of seabirds and are certainly prone to wander around and interfere with nesting birds, amongst which they may be a significant cause of nesting failure.

It may seem incredible that parent frigates would be so unconcerned. Even normally timid birds may be fearless in protecting their young. But the answer may lie partly in the curiously ill-defined territoriality of the frigate, itself a consequence of their peculiar breeding cycle, as I have tried to show. But it is often impossible to interpret the function of behaviour (or its absence) in simple black-and-white terms. We just do not know enough. Even so, the reason for the vigorous interference by 'spare' males remains particularly obscure. It is one thing to suggest why frigates have poorly developed territorial behaviour but (even if this suggestion is correct) quite another to 'explain' why any frigate should attempt to disrupt the nesting attempt of another pair. What do they gain by it? All sorts of possibilities spring to mind, but all are the merest speculation. Could the wreckers be increasing their chances of gaining an experienced mate by ensuring that the females whose cycles they interrupt will be available for pairing next year when they otherwise would not have been? But the disrupted male is also released to compete again! How one would love to have a vast number of marked birds under intensive observation for thirty successive years.

Some fifteen years later Barry Reville, a student of mine, spent two field seasons on Aldabra Atoll in the Indian Ocean, attempting to sort out this and other frigate problems. Although his birds nested quite high in mangroves, he made some fascinating discoveries. He overcame the problem of marking birds by using paint-soaked pellets fired from a blowpipe. In addition, he literally lived in his pylon-hide for days at a time to keep track of the comings and goings. He found that the female great frigates tended to choose the densest clusters of displaying males in which to find a mate. Group after group filled up in sequence. This was not so in the lesser frigates which nested nearby. Their females showed no preference for dense groups, gradually filling up both dense and more-spaced groups simultaneously. One result was that, in the great frigates, late-arriving females tended to have their laying date brought forward as a result of the social stimulation they encountered in the dense groups to which they were attracted. Consequently, egg laying in these groups was more closely synchronised than in those of the lesser frigate. This in turn meant that the period during which great frigate nests were vulnerable to interference from non-breeding males was reduced and breeding success was higher. Shades of Fraser Darling's hypothesis!

The frigates had another puzzling habit. During the hottest parts of the day, when the iguanas sought the shade and everything drooped, frigates sat facing the sun with their great wings spread and turned to catch the heat, or lay with outsprawled wings, presenting their backs to the sun. Often they gaped and panted and looked thoroughly uncomfortable. Many years later Houston showed that vultures behave similarly apparently in order to restore the correct curvature to flight feathers which have been slightly deformed by air pressure during hours of soaring. It was claimed that the heat affected the structure of the keratin of the feathers and restored their aerodynamically efficient shape.

Their aerial acrobatics, quite dizzying to watch, depend

170

partly on their huge vanes and negligible weight but also on the angled shape of the wings and on their long, deeply forked tail. They idle along, high in the sky, gently scissoring their tails, until they spot a booby returning to the island. One would imagine that the advantage would lie with the booby, which has all the initiative in twisting and turning, but the frigate's reactions are so swift and its agility so consummate, that it might as well be the booby's shadow. In vain the harassed booby's desperate jinks, useless its agonised screeches. With maybe half a dozen piratical frigates on its tail, it has no chance of escape unless it can reach the trees – and how greatly it strives. Otherwise, only the jettisoning of its hard-won cargo can save it. Often, the parcel of squid or flying fish are snatched before they even hit the water. But it is a different matter if the encounter takes place on the ground or in bushes. We watched, with some satisfaction, a red-footed booby return to its nesting tree in which two frigates were displaying and send both of them packing. One got its wings so inextricably hooked on branches that we had to rescue it from a lingering death. On the ground the masked booby is even more dominant. But in the air the frigate is peerless. Oddly enough, red-footed boobies robbed of nest material reacted quite differently from those robbed of fish. They screeched loudly and pursued the frigate persistently though futilely – something they never did after losing food. It was another example of different rules applying to different situations, like the gannet that pulverises a herring gull within the gannetry but is thoroughly discomfited by one beyond the edge of the colony. One can't say with any justification that the red-footed booby 'knows' it cannot regain its fish, which the frigate has just swallowed, but 'thinks' it might regain its twig. Nor can it have been successful in the latter endeavour sufficiently often to condition it to try again. It *never* caught the frigate. The probable answer is that the fish was regurgitated by the booby whereas it did not 'regurgitate' or drop the twig but had it forcibly snatched away. Its motivational state was therefore different in the two instances. As I have tried to show, birds are programmed to react in specific ways to specific situations. Sometimes we can see why, in terms of 'pay-off' and sometimes we can't. I can see why a gannet threatens a marauding gull in the colony. I can't see why it doesn't dismiss it equally peremptorily on

The red-tailed tropicbird on Christmas Island.

the fringe of the colony. Clearly, a gannet 'feels' different when it is within the colony (as it is important that it should) and that 'feeling' (motivational state) means that a different set of reactions are available to it. Beyond that, even if so far, we cannot go.

Frigates, boobies and tropicbirds – the big three: the lovely red-billed tropicbird with its long, pendant central tail feathers also bred on Tower, but while the red-footed booby had a limitless number of nest sites, the tropicbirds needed holes and these were not abundant. Fights were common and we found two dead adults, both at the entrance to holes. Yet, in other parts of its range it nests on the ground. Why not on Tower? Later, when we lived on the Indian Ocean Christmas Island, we grew familiar with one of the loveliest of all birds, the golden bosun or tropicbird. Although it is merely a race of the white or yellow-billed bosun, its deep apricot yellow, seen against the brilliant green of the tropical rain forest, sets it far apart. So, too, does its almost unbeliev-able habit of nesting in holes, or hollows between branch and trunk, beneath the jungle canopy. One could hardly imagine a more dangerous habitat for a web-footed seabird which is

172

helpless on the ground. How on earth did the habit evolve? Again, I wonder if the holes in the sea-cliffs, its natural habitat, have been comandeered by its more powerful relative the red-tailed tropicbird, which is the Indian Ocean equivalent of the red-billed. On Ascension, the red-billed often excludes the smaller yellow-billed. Even so, the golden bosun does seem to have gone to extremes. To weave through the leafy caverns of the forest, with trailers, tree-ferns and branches in the way is asking a lot even from an adult, but it is simply preposterous that a newly emerged fledgling, straight from the hole, should be compelled to fight its way through the canopy before it can begin its descent to the sea which may be more than a mile away. Yet such is its fate and, despite the many casualties, enough survive to maintain the population. It is simply another illustration of the fact that so long as the system works well enough, it will do. Evolution does not necessarily produce perfect adaptations – just workable ones. Often, of course, adaptations *are* unbelievably intricate and beautiful, but some creatures are but imperfectly adapted to their way of life.

13

Sea-Lions and Albatrosses

I was on the level sand when he came open-mouth'd
at me out of the water, as quick and fierce as the most
angry dog let loose. I struck the point into his breast
and wounded him all the three times he made at me,
which forced him at last to retire with an ugly noise,
snarling and showing his long teeth at me out of the
water.

Captain Woodes Rogers

If God spewed lava on Tower, I agree with Bishop Berlanga
that he rained stones on Hood Island. Punta Suarez, where
we camped, was not merely littered with stones, it *was*
stones. They were not the wickedly sharp and brittle lava
sheets of Tower but comfortably smooth, age-worn boulders
of a reddish hue. On the south side of the Punta the rusty
cliffs fell onto a deeply fissured terrace pierced with blow-
holes, through which towered booming spumes of spray.
But on the opposite side a delightful sandy beach shelved
away into the clear, cool water. Behind the beach, stretching
away to the centre of the island but thinning out towards the
extreme point, lay grey scrub and cacti, in places totally
impenetrable. We arrived after a twelve-hour sail from Academy Bay and with helping hands had everything unloaded
and the tent up in two hours. This time our stay was to be
shorter and although, like Tower, the island was waterless,
our supply of 130 gallons was enough to last us the four
months we expected to be there.

The beach was thronged with sea-lions and foul with droppings. Heavy bodies had polished the boulders like glass.
Although we were innocently ignorant of the fact, we were
setting up camp in the midst of breeding sea-lions. All day

174

and every day and most of the night sounded the monotonous, blubbing coughs and roars of patrolling bulls. A glance out of the tent showed sea-lions surfing, and cavorting not merely on, but entombed within, the glassy, towering walls of the breaking rollers. At night heavy breathing and snoring betrayed sea-lions snoozing alongside the tent. Heart-stopping roars and crashes accompanied the pell-mell battles of rampaging bulls. More than once, half a ton of sea-lion tripping over the guy ropes threatened to pull the tent down on top of us. One night, goaded too far, I rushed out and belaboured a bemused bull with the first thing I could lay hands on, which happened to be the aluminium legs of a folding table. They bent like soft wax. The morning after a huge intruder bull had roared challengingly just outside the tent, the thought of what might have happened to us if the beachmaster had cornered him there stimulated me to build a barricade, but it was little more than a token.

The focus of the beachmaster's interest was the harem of sleek cows, hauled out and snuff-dry on the warm sand, well beyond the curling, ice-green breakers toppling lazily and thunderously to seaward. To fend off intruder bulls he patrolled ceaselessly just offshore, backwards and forwards in a series of sinuous, porpoising dives. Each time he surfaced he roared, a hoarse bellow after a heavy inhalation like a donkey's bray. There were plenty of intruders for, as in all mammals which are massively polygynous but in which the sex-ratio at birth is more or less equal, the reigning monarch is constantly frustrating other sexually mature males. Eventually these covetous bulls are compelled to make their bid, but it has to be at the right time. The propitious moment for the new bull is when the old one is weak and exhausted by fasting and mating but apparently this juncture was not easily judged. Usually the beachmaster saw off his rival with a single galvanic charge, shoving walls of solid water aside as though they were mere foam. Often they clashed briefly in a maelstrom of churning water before disengaging. Then came the pursuit, two humped, glistening black backs hurtling through the sea.

We did see a few take-overs and knew that others had happened because some beachmasters had recognisable scars. We could easily spot our favourite old tyrant, George, because he had a piece bitten out of his forehead. One morning a new bull, probably younger, for his forehead was flatter than the impressive dome of the master, was patrolling. Around midday, George returned and the new one slipped away to sea but a few hours later it returned and roared whilst George was hauled out, apparently asleep. George awoke and immediately roared repeatedly, utterly silencing the new-comer who lay low in the water and then departed in a series of casual dives. But George himself disappeared soon after-wards. The take-overs were quick affairs rather than desper-ate battles but all the same there was an electric menace about them, a bruising clash of heavyweights. Around mid-day one huge beachmaster was rolling in the shallows when an equally impressive bull approached from the left and reared up, gazing intently at the master. The latter soon looked up and moved menacingly forwards. The new bull approached equally threateningly and they clashed. After a brief fight the new bull, distinguishable by a long, narrow wound on the back of his neck, chased the other out to sea before returning to the beach. Yet, the very next day, the patrolling bull lacked the neck-scar! The following day the scarless bull was itself displaced by a light-coloured speci-men with a circular neck wound and a mere fifteen minutes later along came old narrow neck-scar again and after a short, sharp struggle, took over the beach once more. The next day, lo and behold, he had disappeared again, only to pop up the day after and, like Mohamed Ali, dispel the latest hope-ful. So it appeared that there was a succession of bulls at-tempting to take over and one (narrow scar) dominant for a while and able to go away (perhaps to feed), return and resume command.

Only once did we see a badly mauled challenger and he had a large raw wound on the lower back. Although it looked nasty it may not have been dangerous to an animal of such enormous vitality. Nevertheless he was lying among the bushes the following day and fled precipitately upon the approach of the master. Further along the beach another three hoary-looking bulls bearing slight wounds were also hauled out. When dry they looked much fiercer, with quite savage faces. In this area we found the remains of several enormous bulls and imagined that they had been exhausted individuals, worn out and ready to die, but who knows? If a bull impregnates many females in a short reproductive life it can be as effective as living longer and breeding more slowly, but obviously this can work only if the male contributes nothing to the actual rearing of his offspring. The red-deer stag is a familiar example.

On our part of Punta Suarez the cows did not drop their pups on the beach but amongst a nearby jumble of boulders. They used the beach solely for courting, mating and resting. The bull was a ponderous suitor but the female enlivened proceedings, leading him a merry dance. An oestrous female was recognised by scent but she did not allow the bull to mount her straight away. She eluded him and gambolled blithely along the beach with the fat male in pursuit. A period of play followed, the male nuzzling her particularly in her armpits, a ticklish spot which made her jump. She reared up and bit the skin of his chest and throat. He kept up a muted version of his territorial bellowing and the whole affair seemed thoroughly amiable. When she finally allowed him to mate, in shallow water rather than on dry sand, he just about buried her beneath his enormous bulk. It may have been the sea-lions that broke the camel's back for on 12 August June announced that the old bull got on her nerves, she'd had enough of the daily grind and privations and wished she were home. I was only amazed that she had stuck it without a word of complaint for so long.

Several pups were born in August. The placentae attracted many scavengers including lizards, mocking birds and Gala-pagos hawks. One pup was born near to our beach and became positively affectionate towards us. A second pup appeared but its mother seized it by the scruff and lolloped into the sea, swimming around with the pup beneath the surface most of the time. Only occasionally did its tiny head break surface and we were sure it would drown, but when eventually she did bring it ashore it managed to crawl weakly up to her. Even large pups with well-pregnant mothers continued to suckle. These boisterous youngsters were the bane of the adults' lives. They came rollicking ashore, wet and full of bonhomie and bulldozed their way into the ranks of the dry, blissfully snoozing sea-lions, crawling over them, flopping onto them and nuzzling them until all was turmoil.

The mocking birds were even more trying than sea-lions. The camp followers on Hood were just as bold and innovative as their relatives on Tower, though considerably larger. The

The Hood Island campsite on Punta Suarez (*above*),
with sea-lions in the foreground. (*below*) The other side
of Punta Suarez.

lucky band into whose collective territory we had so provi-
dentially blundered lost no time in investigating this wind-
fall, for simply to survive they have to be relentlessly opportunistic. One of them fell headfirst into a narrow jar half full
of vinegar but, by assuming an extreme head-and-tail-up
posture, managed to avoid drowning until rescued. We dried
it off in the oven and then down June's blouse. For days

afterwards it had to run everywhere because its feathers were stuck together so that it couldn't fly. Another jumped into a bowl of washing-up water, in pursuit of a lava lizard. The forty thieves often frayed our tempers. They repeatedly extinguished the wick of the paraffin fridge and beseiged the tent to such purpose that we dare not leave it without barricading every hole. Even then, just as on Tower, they forced an entry by squirming and wriggling more like mice than birds. In nature, they are supreme scavengers. Does an albatross come ashore to feed its chick? Then a mocker may well attend the operation to benefit from any oil that may be spilt. They managed to scrape it up by putting their head sideways and working their bills against the oil-stained boulder. Albatrosses provided eggs for the mockers for, without obvious reason, they often moved them several feet from where they were laid, losing some in the crevices between boulders. If the mockers could break one open, there was a feast but often they could not manage to do so and the egg lay there until it went rotten. Even sea-lion faeces and decomposing booby chicks were eaten. One of their less attractive habits was to eat live adult boobies. They pecked away at the booby's cloaca until it bled and then delicately sipped the blood. This repellent habit has been taken further by the small-billed groundfinches on some islands, which attack the blood-filled quills of growing booby chicks and can cause serious wounds besides exposing the victims to a nauseating plague of flies.

The faces of mocking birds are individually recognisable. One day, several were crowding around a tin when an evidently dominant bird arrived and simply tossed the others aside like so much litter. In fact it used the same quick head-jerk that is employed in shifting debris. Usually, though, a mocker ran up to another and if its face was hidden it peered round from behind before attacking or running away. Occasionally a subordinate made a mistake and pecked a dominant bird from behind, provoking swift and vigorous retribution. A proper study of mockers would be a delightful project and one that would greatly attract me. Group territorial birds have especially interesting habits and, particularly in arid places, tend to breed co-operatively with 'helpers' who are not themselves the parents of the young. Mockers seem likely to do this.

179

It was not the sea-lions and mocking birds that had brought us to Hood Island, but the blue-footed boobies and the albatrosses, the colonies of which behind the tent were just as handy as the red-footed boobies and frigates on Tower had been. Red-footed boobies were absent, probably because they are warm-water birds and Hood, in the south of the archipelago, is washed by the cold water of the Humboldt current. But this suited the blue-foot which is in fact restricted to the vicinity of cold water both here, in Peru and in the Gulf of California. Luckily, they were setting up nesting territories and courting and we lost little time in getting to work. The blue-foot is a comic. He uses his ridiculously blue feet like flags, with hugely exaggerated, flaunting steps and an athletic landing display which flashes them against his white belly. His head and neck are striped, his eye a piercing yellow and his dagger-like bill a darkish blue. Just as had happened to the red-feet on Tower, so here, within the short space of a month the colony changed from a thriving nursery into a disaster area. There were dead chicks and abandoned eggs everywhere. At first it seems strange to see parents resting and preening at the nest-site whilst their chicks are starving. One might expect the adults to spend every minute scouring the unproductive sea to save their young. But they will produce more young over their lifetime if they avoid too much stress, even if this means letting a few chicks die, especially since most of them will die anyway before they become old enough to breed. The frigates and the red-footed boobies had behaved similarly and for the same good reason.

Some of the adults we caught had been ringed earlier by Miguel Castro and David Snow from the Charles Darwin Station. David had noted the contents of the nests and this lucky break enabled us to calculate that they were breeding every nine or ten months, rather than once a year. Frigates breed less often than once a year and the blue-feet more often, but both strategies are responses to the aseasonal tropical conditions. Since, by and large, one time of year is as good (or, more likely, as poor) as any other, the whole year is available. So cycles of a year-and-a-bit are perfectly feasible. Neither of these non-seasonal strategies would have the slightest chance of working on the Bass Rock but they were appropriate here. This was all exciting and new to me.

It often happens that the implications of fieldwork, itself

A female blue-footed booby with its two dissimilarly sized chicks, the smaller of which starves (but is not actively killed) in times of famine (compare with the masked booby's fratricide, p.160).

apparently trivial, emerge only after analysis of the results. Sometimes, indeed, the significance of the findings depends so heavily on statistical treatment that one may be forgiven for harbouring doubt, but occasionally important pieces of a jigsaw puzzle fall straight into place. In a minor way this happened when we saw a blue-footed booby's nest with two very unequally-sized but well-grown chicks. Obviously the big one had allowed the little one to live although equally clearly it had claimed most of the food. This may seem much the same as happened in the masked boobies of Tower Island, where the older chick killed its younger sibling, but there is a world of difference. By sibling murder the masked booby has evolved a strictly one-chick brood as an invariable policy, to increase its chances of rearing any at all. The blue-foot by contrast gives itself at least the chance of rearing two, or even three. The masked booby has its brood-reduction behaviour, by direct killing, written into its genes. The blue-foot's brood-reduction is not so rigid. If it occurs at all, it does so by exclusion from food rather than by direct murder. The end result, if death occurs, is the same but the 'if' is the important word. The two breeding schemes are very different and the difference can persist because in the blue-foot's case, but not the masked booby's, it is sufficiently often practicable to rear two healthy youngsters. Another bit of jigsaw fell into place when we saw, with some surprise, that the male blue-foot was tiny compared with the female. Not only (because of their proximity to more productive cold water) were the blue-footed boobies less prone to forage far out at sea than the masked, but there was clearly a marked difference between the sexes with respect to feeding niche. The small, agile male was hardly likely to feed in exactly the same way, and on the same-sized prey and in the same places as the powerful female. In fact we soon discovered that the perky little male was an amazing acrobat. One, which seemingly hurled itself to destruction by diving onto the rocky foreshore had in fact dived headfirst into a small rock pool less than two feet deep. A female blue-foot couldn't have done that.

So far as brood size is concerned this means that the blue-foot chicks, which for the first two or three weeks need to be fed fairly frequently, have the advantage of parents that feed nearer home and in particular of a father who can make short, quick trips, even if he doesn't bring much back each time. Two chicks are therefore a better proposition for this species than for the masked booby. If hard times do come, then naturally the bigger chick claims the lion's share and the younger one may die.

Trick dives weren't the only accomplishments of the blue-foot. On one occasion we saw a colour-ringed male that we knew to be the parent of a partly-grown chick, displaying to over-flying females more than fifty yards away from its chick-site. Ten minutes later he was back with his chick. Unfortunately our time was up before we could discover whether he began a new breeding cycle with a new female whilst his old female was left to rear the part-grown chick on her own. This is perfectly possible but it would have been a novel discovery at that time.

Enough of boobies! The waved albatrosses were too interesting to be ignored. Like the fur-seals and the penguins, the albatrosses had come to the Galapagos via the cold-water Humboldt current. Whilst not so spectacular as the enormous royal and wandering albatrosses they were still impressive, with that dark, gentle albatross eye, a great, hooked yellow beak and a finely vermiculated breast which gives them their name. Many of them had downy youngsters when we arrived – dark chocolate-coloured chicks, delightful when small and clean, unbelievably ugly when large and bloated with oil, their beautiful down soiled and spiky. They became extremely heavy, weighing more than 5 kg, and could not at that stage stand upright for more than a minute or two but had to squat on their tarsi with their great bellies bulging forward. One felt that they needed a smooth, soft sward with a few friendly bushes for shade. They could then have settled their paunches gently into a convenient hollow and snoozed until the next feed. On Hood, they had boulders to trip over, innumerable and inescapable, and thorny scrub to entangle them like sheep in a briar patch. Moreover, along the north coast they actually chose to nest among dense scrub, through which the adults were forced to thread a tortuous course, in some cases for at least two hundred yards from their cliff-edge landing place. Yet there were innumerable lovely clearings fringed with bushes for cover and perfect for landing, devoid of albatrosses or with just the odd bird.

Sometimes chicks which were attacked by adults stumbled and fell whilst trying to escape. This was particularly dangerous near the edge of the cliff, for they obviously lacked the fear of edges which genuinely cliff-adapted species show. Walking at some distance past one large chick we were horrified when it shuffled to the edge, stood uncertainly for a

183

184

185

Components in the ritualised 'dance' of the waved albatross (*two previous pages and above*).

moment and then, without any further movement from us, stepped over the edge. Perhaps it is the absence of a cliff-response which makes the apparently perfect cliff-top strip unsuitable for albatrosses.

Even though many pairs were busy with chicks, some displaying birds still remained – perhaps pre-breeders in the main, for albatrosses may well need a year together before they even try to breed. These carefree birds gathered in groups beneath the sun-silvered scrub, to emerge in the mornings or late afternoons and begin dancing. Both partners played an active role, whereas in some displays, including those of many birds, it is the male alone who performs; the female merely watches. The crested newt, which I have watched for many scores of hours, is an extreme example of this type. The male goes through a complicated series of actions rather like a washing machine which proceeds from fill to wash, rinse and spin. All the washer needs is a constant source of energy. If that is supplied, it goes through its pre-set routine. Similarly, all the male newt requires is a receptive female and he will go through his routine. She crouches on the bottom of the pond and he goes through an elaborate display in which one component follows another in a predictable sequence. In other courtships, such as that of the gannet's mutual fencing display, both partners participate but they both do basically the same thing at the same time as though two washing machines were working synchronously. But the dance of the albatross was like two washing machines, each with the same repertoire but out-of-phase with each other. Yet not randomly out-of-phase. The spin of washer B was more likely than any other part of its repertoire, to follow the rinse of washer A.

By simple shorthand and with the help of a stopwatch, I dictated the timed sequences of behaviour whilst June scribbled them down. The dancing was fast, furious and vocal, each bird snapping into and out of its various postures with whinnies, beak-clappering and loud clunking noises produced by the amplification of beak-shutting. How the amplification was produced remained a mystery – maybe the buccal cavity acted like a sounding chamber but whatever the mechanism the result was a resounding 'clunk'. From the adults that we caught and colour-ringed we discovered that groups formed and increased over a period of several days, individuals dropping in and remaining a variable time before leaving – rather like a hippie commune. Most of the pairs which formed and danced during those shore-leave periods were not constant, again like a commune, which suggests that it was not so much the formation of a durable pair-bond but the dance itself that was the object of the exercise. Our crying need was to know the age and status of the dancing birds but this information was, alas, completely beyond our reach. Later, however, Mike Harris showed that these albatrosses require at least one season's display together before breeding for the first time.

One day we managed to see the entire process by which the waved albatross feeds its chick, from the landing of the adult, through the location of its own chick to the final transfer of oil, as much as 450 g in each of perhaps twelve separate feeds and all in less than five minutes. The parent 'homes' onto its chick, which, along with others, may have been hidden anywhere in a wide expanse of scrub. The adult called continuously as it waddled forwards and the chick called in response. The ability to recognise each other's voice did the rest. Once contact had been established, the oil-tanker and the receptacle connected up. The chick inserted its slightly open beak crosswise into that of its parent, who stood or squatted with lowered head and pumped oil and partly-digested food from its stomach and crop into the trough of the chick's lower mandible. Very little was spilt and that gratefully gleaned by those oil-stained scavengers, the mocking birds. The chick's sagging belly swelled visibly as the oil hosed down into it. After such a colossal feed it staggered unsteadily back to its shelter beneath the scrub, where it collapsed in a heap of moth-eaten brown down.

The cold waters around Hood produced the *garua* (mist) which cut out the sun's heat and produced dull, grey weather much too miserable to go without clothes. The north side seemed even colder and June, who had begun to think like a bird, said, quite without pretence, 'I'm glad *we* nest on the south side.' By 17 September we opened the last of our tins of sausages. At this stage we would have refused the opportunity to remain for another year, even with new provisions. This may seem weak, for there were many things still to be done, but we had simply had enough. June calculated that we had carried out 3,216 weighings, to say nothing of measurements of bills, wings, examinations for moult and all the other work. But we still did not laze around. As late as 12 October, between us, we filmed, carried out an egg check in the colony of blue-footed boobies, weighed and measured several albatrosses, chicks and adults, typed part of a manuscript and cooked. But by 9 October we had begun, happily, to pack. Box number 4 containing most of our books, blankets, hide, clothes and specimens was finished first. Four days later we were startled to see the unmistakable rig of *Beagle II* on the horizon. Apart from a brief visit by a fishing boat this was our first contact with anybody since our arrival. On board was the station's new director, Roger Perry, young, ex-Cambridge and BBC, not a scientist but extremely likeable, diplomatic and an excellent administrator. We were invited on board for supper – alas it was tuna, our staple diet all year – but their news was more exciting; the Duke of Edinburgh was to visit Hood in the *Britannia* in just over a fortnight's time. By then, as it happened, we were down to half a tin of margarine, one of Spam, some tuna and some rice, flour and jam – not bad reckoning by June almost a year earlier. Paraffin was low, chairs broken and clothes in rags.

On 4 November, a day which was to prove an ideal contrast to the whole of the past year, the *Britannia* slid neatly round Punta Suarez at 8.30 a.m. and made a perfect rendezvous with the *Beagle*, hastening in from the north. Almost before the propellors had stopped two rubber dinghies were on the move, one to the *Beagle* and the other ashore. The Duke's party included Aubrey Buxton, whose grandfather reintroduced capercaillie to Scotland towards the end of the last century. Aubrey is a keen ornithologist and conservationist and has done a great deal for the Galapagos.

However egalitarian one may be (and I'm not, particularly) one cannot pretend that there is no difference between the Duke of Edinburgh and Joe Bloggs. I have described the occasion in an earlier book, but lunch on the *Britannia* was so incongruous that I feel bound to repeat it here. We had expected nothing more than to show the party a few birds and give them the chance to take pictures of a few special subjects, for HRH was a keen nature photographer, so the invitation, which was quite spontaneous, came as a complete surprise. June looked alright, but I took my seat in the dining room, along with the Admirals, with bare feet and patched and ragged shorts covered with albatross vomit. It was not unlike the archetypal dream in which one walks along a crowded street with nothing on.

The *Britannia's* carpets, I recall, were dove grey, the woodwork softly gleaming and the lighting subdued. For half an hour, Cabin 12 was ours, with bathroom, lounge, writing tables and easy chairs. The panelled corridors were hung with pictures of ships and yachts. On the dining room table, amidst the silverware, stood a huge arrangement of irises, quite up to the standard of our own table decorations these twelve months past. How better could we have rounded off our Galapagos venture?

I am a great fan of Prince Philip's (and, latterly, of Charles). This is not just because of his position but because he is an exceptionally concerned person. As everybody knows, his interests and abilities range widely but, if those to do with conservation are any guide, always with depth, insight and great effect. He has certainly done a lot to save Christmas Island (chapter 15). And he is so natural. It was typical of him that he offered to take all my films and notebooks back to London to cut out the risk of loss during transit. When, some years later, we met a couple of old ladies whose Morris Minor had slid off the single-track west highland road into a ditch, Prince Philip didn't just ask us all to get out and lift it back onto the road – he jumped out and did it himself and not for effect; it went quite unrecorded. Whether the shock to the ladies did more to shorten their lives than the accident had done is beside the point.

The day after the *Britannia* pulled away from Hood we left in the *Beagle* but before quitting the Galapagos we made another trip to Tower in late November, mainly to check the state of breeding of the various seabirds. 'Poppet', our lovely little finch, met us when we landed and went through his usual finger-nipping routine. We were greatly impressed that he had evidently recognised us after five months, though I suppose it is possible that he treated all humans alike. The frigates had done abysmally badly and hardly any young had survived. This was partly the result of leaving small chicks unguarded whilst they were still vulnerable to intruding frigates and to short-eared owls. These fierce little predators killed not only young frigates but well-grown booby chicks, by attacking the back of the neck. As I mentioned, they wreak havoc among the storm petrels and I once disturbed one from the carcase of a freshly killed whimbrel. It had eaten both eyes and the skin from the head.

The colony of Madeiran and Galapagos storm petrels on the horn of Darwin's Bay was buzzing with activity just as it had been seven months previously. Presumably there were successive waves of breeding birds or, less likely, a constant attendance of vast numbers of pre-breeders.

At 5.45 a.m. on 27 November 1964 we finally left Tower. All the excitement of our landing the previous December was absent – it was simply a departure; as an event, it was quite flat. I didn't even leave my bunk to see it recede into the greyness. But many a time I dream of that little beach and the lagoon. Once out of Darwin Bay it was quite rough. At one point water spurted through the planking into Carl's bunk, bringing him on deck in double quick time. As it happened, the old DC6 in which we were hoping to fly from Baltra was full of air-force personnel travelling to the mainland for Christmas and we were lucky to be allowed to join them. Weight, apparently, was critical. Everybody was frantically weighing things – dried fish, lobsters and, of all things, crates of lava rocks. Goodness knows what they were for. Our luggage was theoretically restricted to a tight fifty pounds and we went to enormous and quite unnecessary lengths to pare down our belongings. Nobody checked. Everybody piled aboard carrying 'set' and varnished lobsters which they hung from the uncovered electric wires running along the fuselage. I was the only one without a parachute and June and I, alone, refrained from crossing ourselves before boarding. The pilot had landed in three giant hops and although he didn't take off in like fashion, he only just hoisted her into the air before the cliff edge. However, the wrinkled sea crawled slowly past beneath us, and at least we were heading for the mainland, hard to miss.

On arrival in Guayaquil we homed straight back to the old Pension Helbig where we were given a much nicer room than last time. Within an hour we had eaten and were taking a rest. Outside, the stultifying heat was pierced by the noise of traffic, horns blaring, radios, the babble and muted roar of a bustling city. Through it I could hear the weird call of the swallow-tail and the courtship braying of the albatross. But already they were fading. The past had gone, the future may never come. The turmoil of Guayaquil was now the reality. When we walked out of the Helbig, who knew that we had just come from a desert island? And what did *we* know of other people's past? Not only on this occasion but many a time since, and as others have noted, the end of an adventure has seemed an anticlimax. But this was *not quite the end,* for to the south lay some of the most awe-inspiring seabird islands in the world, the Peruvian guano islands, the home of millions of boobies, cormorants, pelicans, terns, diving petrels and penguins. We were too near to miss the opportunity.

14

The Marvellous Guano Islands of Peru

I have seen the penguin regiments of the far south,
the courtship antics of the Wandering Albatross, a
file of eighteen Condors which passed me within a
stone's toss, and other marvellous sights in which
birds held the centre of the stage, but nothing more
exquisite than the pantomime of adult piqueros hanging on the wind above the cliffs of Guañape.

Robert Cushman Murphy

It was mid-morning, Friday 18 December 1964. The twenty-
five-year-old *Pacific Queen*, a beamy sixty-footer, wallowed
in an oily, turquoise-green swell outside the Peruvian port of
Callao on her way to the legendary Guano Islands. By the
greatest of good fortune, we were aboard. To this day I do
not know why the illustrious Compañia Administrado del
Guano had paid the slightest attention to an obviously impecunious and unimportant bird-watcher and his equally
obscure wife. But attention, dignified and ineffably Hispano-
American, they had paid, to such effect that we were now
their guests, collected by taxi, conveyed in comfort, with our
own cabin and to be hosted on the island by the impoverished, but impeccably courteous and generous, guardianes,
or keepers.

The harbour was crowded with anchovy fishing vessels
exploiting the trillions of small, sardine-like fish that swarm
in the cold, rich waters of the Humboldt current. The foot-
hills of the Andes, already indistinct in the mist, sloped
down to the calm ocean which, here, deserved its name, 'Pacific'. There were seabirds everywhere. Fantastically adorned
Inca terns with blood-red bills and white ear-crescents curv-
ing out from their heads, lumbering Chilean pelicans, busy

193

cormorants, boobies, gulls, skuas, shearwaters. I'd never seen anything like it and this was only the curtain-raiser.

The voyage took thirty hours – more than a day cruising slowly down one of the most fabled seabird lanes in the world. And at the end of it, two islands, Guañape Norte and Guañape Sur, holding more seabirds on their barren slopes than most of the British seabird islands put together. On the deckhouse walls hung fresh meat and vegetables; on the deck two comfortable chairs and in the chairs two exultant travellers. In our cabin, clean bunks. On the table, good plain food. Outside an endless procession of seabirds – ropes of them, skeins, thick, undulating lines, flocks, multitudes. What more could one desire? At that moment, nothing. It was, in its way, complete, better by far than a Lindblad cruise.

We arrived at Guañape Norte at 1.30 p.m. for a mere three-minute stop. These islands were utterly barren. The trite phrase 'not a blade of grass or any other vegetation grew' was literally true here. There was guano, rock, birds and concrete; nothing else. The buildings, quite large, were nearly empty – several bedrooms with no beds, a bathroom and a bare living room. The four guardianes lived in dirty, ramshackle huts. We were in the guests' quarters. Everything and everybody was smothered in fine dust from the guano. And the birds! The island was a patchwork quilt with guanay cormorants, like pile on a carpet, forming discrete black masses and boobies, slightly more spaced, lighter patches between them. Some areas recently worked for guano were entirely deserted, bare ochre or cream-coloured rock. Ten thousand, perhaps twenty thousand birds were in the air. More thousands rafted on the sea. When dusk came, unbroken streams poured back onto the rock in thick, endless ropes. In the tumult of a hundred thousand raucous, gabbling seabird voices all individual sounds were lost. The roaring, rising and falling, remained as a background, pervasive as the dust. Perhaps only the most populous of the great northern seabird bazaars, at their height, or the teeming colonies of the Antarctic, could remotely compare with this spectacle.

We landed on the northern Guañape Island but the following day we accompanied the guardianes across the channel separating it from the southern Guañape. We embarked in

Piqueros (Peruvian boobies) on the guano island of Guañape Sur. (*above*)

The sexual advertising display (skypointing) of the piquero. (*below*)

enveloping mist, the *nublita,* and ran into a vast gathering of cormorants which began to patter over the surface before we could see them – a weird effect like muted applause from a vast, unseen audience. Soon they circled over the boat in thousands and we were in the centre of a moving cloud. Some Humboldt penguins added their voices, like cows wanting a bull – most unbirdlike. The southern Guañape was inhabited by immense numbers of piqueros (Peruvian boobies) – I reckoned more than a million and the guardianes independently suggested the same figure. Many nests held two or three chicks, plump youngsters in their nests of grey, concrete-like guano. There were no signs of starvation here. But when starvation does come, it devastates the entire population. Robert Cushman Murphy, whom I mentioned earlier, has bequeathed memorable accounts of these islands and their birds. The *niño* phenomenon has been well documented for almost half a century and was recorded for hundreds of years before that. Essentially, it is a failure of the cold-water upwelling whose nutrients sustain the food-web that culminates in the anchovies and their predators. It cuts off food at source and for the duration of the phenomenon there is nothing the birds can do except starve or emigrate. Most of them starve. It isn't comparable to the sudden scarcities that I have described for Galapagos waters because there the birds can simply forage more widely and the adults generally do not starve even if the chicks do. But at the time of our visit the Peruvian colonies were thriving.

Christmas Eve saw us well-settled in our spartan quarters, June knitting by the feeble light of a hurricane lamp. Our sacking-covered beds, festooned with cobwebs and guano, were infested with mites from the birds. Beneath the building, the labyrinthine sea-caverns gurgled, slapped and soughed. Outside, the *nublita* enveloped the island. On this night, in my boyhood, the ginger wine would have been steaming by the coal-fired range, like Mr Micawber's fragrant punch, in a vast yellow-lined earthenware bowl, the Christmas puddings standing in their cloth-covered basins, the mince pies a-making and the Christmas fairy gracing the spire of the fir-tree. Not that I would have swapped. Here, our Christmas fare was lumpsuckers, which cling onto vertical rock surfaces in the tidal zone and were eagerly sought by the guardianes although their glutinous texture was not to

196

our taste. The kitchen was unbelievably sooty and cob-webbed and filled with smoke from the smouldering stove. Rowing back in the beautiful last light the sea was calm, the island-tops clear in the sun and tens of thousands of boobies were diving into the shoals of anchovies. Yet, after a week or two at home, all this would seem infinitely remote and we would recall mainly the dirty food, the bugs and the smell. The truth is even sadder – both the beauty and the squalor would diminish and the memory become a mere wraith, a miserable shadow of the event.

So Christmas day dawned among the guano birds. I had often hoped that we would be amongst these teeming colo-nies and here we were. But the flesh dies hard and lumpsuck-ers, amidst the soot and cobweb of a corrugated iron cooking hut, somehow fell short of a plump goose or a fine leg of pork. Still, there were years of pork ahead, if that is so important. Meantime we had but to stroll through our door-way and we were amongst seabirds by the million. The guano keeper of this island reckoned there were 356,346 pairs of piqueros here. I wouldn't vouch for the odd 46 pairs but there were certainly hundreds of thousands of birds. He classified them as 'dense' at four pairs per square metre. However, the cormorant, or Guañay, considered this to be colossally extravagant and almost a million packed them-selves in at seven pairs per square metre.

In the twenty years between then and now, these marvel-lous seabird cities have been thoroughly devastated. They had always been subject to spectacular mortality when the anchovies failed, as they periodically did. But to these natural catastrophes brought about by the failure of the cold upwel-ling and the consequent movement of the cold-loving an-chovies into the deeper layers where they are innaccessible to the birds, recent years have added the morbid effect of mas-sive overfishing by man. It is difficult to remain calm when writing about this sort of thing. The senselessness of the onslaught is staggering. Despite all the signs and all the warnings from expert bodies, the astronomical toll taken from the fish population continued. When the birds were then hit by a *niño* the depleted anchovy stocks were unable to fuel the usual rapid recovery with the result that, today, the guano birds that formerly numbered between twenty and thirty million, and regularly climbed back to those heights

197

after their 'crashes', remain at about a tenth of that figure.

The guano birds of the Humboldt current clearly show that food can regulate a population of birds without competition for that food entering into the matter. In normal times, before the bird population crashed, food was superabundant; there was far more than they could possibly eat. Thus, when commercial fishing began, the catch, totalling millions of tons, was removed from the fish stocks without affecting the birds. If the birds had been competing for the fish, the removal of millions of tons of such a limiting resource would have made it much harder for them to cope and many would have died. But they didn't. Those millions of tons were 'extra' to the birds' requirements. When the *niño* struck and the anchovies became largely unobtainable, then of course the birds starved. Food therefore limited their numbers but it did so independently of the size of the population that existed before and during the shortage. In other words the size of the bird population was not the cause of the crash.

However, booby behaviour was my immediate interest on the island. It struck me at once that the piqueros were intriguingly like the blue-footed boobies we had just left behind on Hood. In fact the resemblances were particularly compelling because the piquero and the blue-foot were clearly very distinct species – there was no possibility that they were mere races or sub-species. Indeed, they even nested side by side on one or two of the islands. There were also striking ecological differences between them and one could begin to see the dim outlines of their evolutionary past. It is, alas, true that the construction of 'just-so' stories is a beguiling pastime. It has the appeal of the chase combined with the thrill of discovery and, for some, even the aesthetic satisfaction of contemplating a masterpiece – an evolutionary master-piece. So one has to acknowledge the dangers of unbridled speculation. On the other hand, a great deal may be learnt by constructing well-based scenarios. In the case of the blue-foot and the piquero the first step is to look at the similarities and differences in their physical appearance, ecology and behaviour and then to suggest some of the implications, perhaps testable, perhaps not. Then, when a reasonably distinct picture has emerged, one can stand back and look at it critically.

Physically, the piquero lacks the spiky, chrysanthemum

head and gaudy ultramarine webs of the blue-foot and where-as the male blue-foot is tiny with an exceptionally long tail whilst the female is large, both the male and female piquero are medium sized, without the noticeably long tail.

Ecologically, one important difference is that whereas the blue-foot breeds in several areas of the tropical eastern Pacific the piquero is firmly restricted to Peru and northern Chile, on the doorstep of the rich Humboldt current. Another difference, not unrelated to the first, is that the piquero nests much more densely than the blue-foot and uses cliffs, which the latter very rarely does.

Behaviourally, territorial fights are far commoner in the piquero than in the blue-foot, and some of the blue-foot's most conspicuous ritualised behaviour is missing from the piquero's repertoire. There are also several subtle differences between them in their pair-bonding behaviour.

One could, of course, simply document all these facts in appropriate detail and leave it at that, but such an offering is not totally unlike giving someone a telephone directory to read for interest. We can at least try to go a little further.

The pronounced size difference between the sexes in the blue-foot shows that male and female are specialised for different fishing techniques and equipped to catch somewhat different prey. Just as the cock sparrowhawk deals with smaller passerines than his larger mate, so the male blue-foot couldn't handle the large fish which the female can. But aided by his lightness, and his long tail, he can dive full tilt into extremely shallow water. Such diversification must open up more feeding niches which, in the areas inhabited by the blue-foot must be advantageous. Thus, in the Galapagos for example, it must partially buffer the effects of the sudden food shortages which occur from time to time. The piquero's case is very different. Its prey is almost exclusively the fairly small anchovy which, as I have already stressed, normally occurs in unimaginable abundance near to the breeding colonies. So there would be little feeding advantage in a marked size difference between the sexes and no need to specialise in different prey.

Of course, pre-eminent though it may be, feeding behaviour is not the sole determinant of physique. Nesting requirements are also important. Here again the differences between piquero and blue-foot make obvious sense. The

199

piquero is, or was, desperately crowded. Those barren rocks, glistening with icing sugar in the hazy blue sky, are so advantageously situated that space on them is the subject of intense competition, not only within but between species. The seabirds cluster on its slopes, pack the broad ledges, cling to the tiniest protruberances, enter the darkest gullies and caves and, in the case of the diving petrels, even burrow beneath the crust – anywhere so long as they can manage to rear their young. In the face of this competition the piquero has taken to ledges as well as slopes and has accepted extremely high nesting densities. Even so, it often has to fight to gain the necessary space. So overt fighting, and the next highest form of aggression, which is the lunge, or threat with open beak, are common. Indeed, as happens in the Atlantic gannet, both partners defend the site almost equally and, even within their own pair-relationship, aggression is a clear and intrusive force.

Now take the blue-foot, dotted around its much less crowded colonies on flat ground or gentle slopes. Space is not at anything like such a premium and overt aggression is far less common. There is less need for shared defence of the territory – indeed, sometimes the male holds two territories at the same time – so the marked disparity in size between male and female does not reduce their ability to defend their site. Even some of the differences between piquero and blue-foot in ritualised display seem easy to understand. Thus, the blue-foot's early courtship involves repeatedly circling the nesting colony prior to landing on his site with a dramatic, skywards up-flinging of his gaudy blue webs, which flash against his white belly. In this aerobatic, he is well aided by his slight build and long tail. One might have supposed it was simply a braking manoeuvre were it not so exaggerated and restricted to the male at one short phase of the nesting cycle. After all, they have to land there later in the cycle but manage perfectly well without this gesture. The piquero doesn't attempt this salute. For one thing it would be suicidal when coming in to land on a crowded cliff face especially since he is bigger and less manoeuvrable than his congener.

There are many more differences with which I could bore you, but where does it all take us? First, it must be said that not *all* characters slot so neatly into place. For example I have no idea why the blue-foot has a streaky head and the

200

piquero a white one, or why the blue-foot's eye is yellow whereas that of the piquero is ruby-red. But I have no doubt that these attributes *do* have a function. It is simply that some differences are more obvious than others, to the simple mind of man. This is an important position to take, even if it is not a new one, for it asserts the all-pervasiveness of natural selection as a mechanism in evolution, a concept which is by no means unchallenged.

Second, these differences, however intriguing, do not tell us whether the ancestral sulid, which gave rise to the two species, was more like the present-day blue-foot or the present-day piquero. That they *did* share a common ancestor is beyond doubt, but what was it like? Well, for what it may be worth, I would speculate that the piquero has diverged from the ancestral sulid rather more than has the blue-foot, in becoming adapted to the highly specialised conditions of the Humboldt environs.

Our spell on Guañape came to an end abruptly and all too soon. The *Pacific Queen* arrived unexpectedly on 29 December – five days early – and whisked us off to the evil-smelling Chimbote, whose fish-meal factories were converting the lovely, nutritious anchovies into chicken food for fat North Americans, whilst hundreds of thousands of Peruvians were starved of the very protein that the fish would have provided. I have no doubt that the politicians could tell us why it must inevitably come to pass that all the sensible things in life are, alas, impracticable.

We returned to Lima just in time to see the city in its annual litter-snowstorm, when at midday on New Year's Eve, all the offices tear up their old calendars and throw the pieces out of the windows. As the capricious papers swirled through the air, they were so many seabirds, closing the most memorable year of my life, an unrepeatable phase worth even more than I knew.

15

Christmas Island and Abbott's Booby

> That the land is a community is a basic concept of
> ecology, but that land is to be loved and respected is
> an extension of ethics. We abuse the land because we
> regard it as a commodity belonging to us. When we
> see the land as a community to which we belong we
> may begin to see it with love and respect.
> Aldo Leopold

The Bass Rock gannets led to the Galapagos because in those
enchanted isles nest three of the gannet's tropical relatives.
The Galapagos pointed to the teeming piquero colonies of
Peru. There remained only one spot on earth where the
world's other two boobies could be found and early in 1967
we packed our old wooden boxes once again, this time for the
Indian Ocean Christmas Island, not to be confused with the
Pacific Ocean atoll of the same name, where nuclear tests
were conducted by Britain. I had no idea that I was about to
begin an affair which lasts to this day and has cost me more
miles, more paperwork, more headaches and more conservation politicking than anything else I have done. Simply the
names, Christmas Island and Abbott's booby, I find pecu-
liarly disturbing.

Once more it was a shoestring expedition, just the two of
us and a couple of boxes of gear including fifty feet of rope
ladder. I knew next to nothing about the mysterious Abbott's
booby, and nobody else knew much more except the late
Gibson-Hill, a gifted naturalist and medical doctor. But I did
realise that it nested high in the jungle canopy and I thought
a rope ladder might prove easier than climbing irons. Much
later, as I dangled in mid-air, far below even the first of my
target branches, it dawned on me that free-swinging rope

202

ladders, especially if they stretch, are fit only for spider monkeys.

We left Southampton on the old P & O liner *Orsova* on 8 February 1967. The channel was hostile and wintry. An adult gannet cruised smoothly alongside though we were travelling at twenty knots in the teeth of a fierce wind. Goodbye Solan Goose – we're off to warmer climes. After a nasty force ten in Biscay, Gibraltar came up, plastered with pro-British slogans and tame baboons. After that, as we slid deeply into the tropics, the magic of a good old-fashioned sea-passage asserted itself. Deck cricket on coconut matting was a keen contest between passengers and crew, and the poolside buffets were real table-groaners. How can one compare that leisurely acclimatisation with the crude shock of arriving jet-lagged after a poisonous incarceration? Quicker it may be; progress it is not. We actually sailed past Christmas Island but the Captain declined to put us ashore so we had to retrace our steps in a dusty old phosphate boat, *The Truth*, out of Bunbury in south-west Australia.

We landed, as all ships must, at Flying Fish Cove, named in 1886 after a survey vessel of that name. It isn't particularly sheltered but quickly shelves to great depths, which is useful for mooring large vessels. It was here that in 1888 Andrew Clunies-Ross brought his party of Malays when he settled the island prior to the first attempt to dig for phosphate, which had recently been discovered. Some of the coconut palms which they planted still survive. Poor old Christmas Island. Its inevitable demise had begun. For more than two hundred and fifty years after its discovery in 1643 this gloriously forested island had remained untouched, hiding its treasure beneath a dense green mantle. Whilst the Galapagos Islands were being plundered wholesale by whalers and buccaneers and the Peruvian guano islands stripped of their immense caps of guano, Christmas Island remained virtually unknown, a remote island of jagged coralline limestone, raised in terraces above the rim of a shattered volcano in the deep waters of the Indian Ocean. Its endemic wildlife included a booby, a frigatebird and a bosun – all of them inaccessible and unknown. Even in 1967, man's impact had been merely local, apparent rather than real. Despite the sophisticated settlement, more than ninety per cent of the jungle remained intact, most of it untrodden. Although it

203

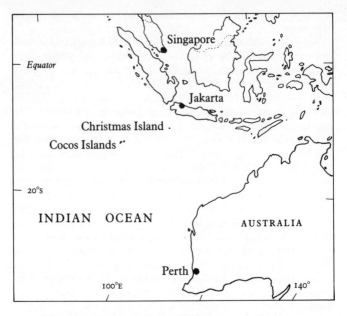

Christmas Island, in the Indian Ocean.

had been visited by the occasional ship in search of fresh
water and food, before its settlement the island's formidable
defences had repelled all intruders. After the terraces of
jagged limestone, backed by thickets of saw-edged pandanus, had been breached, the way to the summit was barred
by an inland cliff, often culminating in treacherous crevasses
draped in thorny trailers and curtains of roots. It cost days of
wearisome toil, hacking foot by foot, to climb the slope
above Flying Fish Cove, where now the Toyotas sweep in a
couple of minutes. It was not until 1887, more than two
hundred years after its discovery, that members of HMS
Egeria cut their way to the top. The following year, on 6
June, it was annexed by Britain.

For the first ten years after settlement there were never
more than twenty-two people on the island. Then in August
of 1896 Murray and Clunies-Ross, who had been granted a
joint mining lease five years previously, made it over to a
company to be formed and registered in London under the

name of The Christmas Island Phosphate Company Ltd. Australia and New Zealand bought the Island in 1948 and it is only within the last five years that Murray Hill, the highest point on the island at 1,114 feet, has felt the shattering impact of the bulldozer and now stands stripped and forlorn after all these centuries. The first consignment of the valuable dust that was to be the island's undoing was loaded, by lighter, in 1900, a suitable act to introduce the century which has devastated the earth.

The year we went to Christmas Island, 1967, was a watershed in its history. Momentous changes were looming. At that time the Phosphate Commissioners were Lords of the island. No feudal Baron held greater sway over the lives and fortunes of his people than did the Commissioners over Christmas Island. But, if it was a dictatorship, it was benevolent. Life on the island was pleasant for all – yes, all – and idyllic for some. The Commissioners, in their pursuit of phosphate, had the legal right to do as they pleased with this glorious island – not for profit but for fertiliser to be sold at cost to the farmers of Australia and New Zealand. There were few pleasanter places on earth, especially for the Europeans. The manager's spacious office, graced by a portrait of the Queen, exuded solid, old-fashioned dignity. Things were done properly. And they were done in style all over the island, for the Commissioners' writ extended to the hospital amidst its well-watered lawns, to the homely little post office, to the school set on the hill above the sparkling sea, to the social club and, most unbelievably, to the carefully tended golf course, alongside which nested one of the world's rarest seabirds, the Christmas Island frigatebird. The gracious bungalows with their verandahs, riotous tropical shrubs and cosseted lawns, the swimming pools, the well-stocked store, the outdoor cinemas, the free bus service – all this and more came under the writ of the Phosphate Commissioners. For single European men, breakfast in the mess offered cereals, chops, kidneys, bacon and eggs, toast and jams – any or all. The Representative of the Australian Government, which in those days was a very minor force on the island, lived in simple elegance in the Official Residence, splendidly situated on a headland beneath soaring crags. The long drive skirted the cove and passed near to a huge boulder with a tree sprouting from a crack, in which a bosun bird nested. The

The settlement in Flying Fish Cove on Christmas Island, showing the phosphate pipeline and the Malay campong.

pleasant little police station, untroubled by crime, was in the charge of Mr Farnsworth, a solid and dependable Englishman. My Christmas Island driving licence was signed, as was usual, with a thumbprint. An ancient jeep and all the petrol I wanted was provided, free. Does it sound too good to be true? I'm afraid it *was* too good to last in the rapidly changing post-war era, though with the exception of two years' occupation by the Japanese during the war it had already lasted for more than half a century. The workforce, some 3,000 Chinese and Malays, were largely content, whatever the present union leaders, a new breed and newly on the scene, may say. Many had lived on the island for thirty or forty years and their children after them, although they depended on the paternalistic goodwill of the Commissioners, and were paid low wages. The sour and divisive atmosphere that emerged later, and exists today, came with the overly milit-ant unionisation of the workforce and the re-organisation of the administration. This is not to deny that some change was both inevitable and desirable, for there was undoubtedly exploitation, but those of us who knew the island both then and now have no doubt that the social atmosphere, across the board, has changed for the worse. However, perhaps I shouldn't comment on political matters that have nothing to do with me.

Life for the Abbott's booby, also, was much as it had always been. Things were still the same as when Gibson-Hill, armed merely with a barometer and his rucksack, set off for days at a time in the trackless jungle. Phosphate was mined in two areas of the island, north and south, connected by a narrow-gauge railway. But in 1967 the entire island was surveyed for phosphate, using a grid of drill-lines, and it was planned to increase production to three million tons per year. Three million tons is a lot of phosphate. This was the watershed. We were the first and the last to study Abbott's booby before jungle-clearing greatly affected the island. At that time nobody had even seen an egg or a juvenile, although at least one identification manual featured a figment of the imagination purporting to be a juvenile Abbott's. Nobody knew how many boobies existed (the most recent guess suggested less than a hundred pairs) and its ecology and behaviour were a complete blank. It was a mystery bird. By now, I must confess, I felt a degree of proprietorship over gannets and boobies – they were my particular domain and Abbott's the greatest attraction of all. It isn't often that one has the opportunity to put what amounts to a new species on the ornithological map, especially a seabird as large and beautiful as Abbott's. It seemed as though it had been wait-ing for us.

The initial discovery of Abbott's booby as a new species,

as recently as 1892, is something of a puzzle. The American naturalist W. L. Abbott collected a specimen from Assumption Island in the western Indian Ocean but he didn't realise it was a new species. Moreover he may have put the wrong locality on the label, the booby possibly having been shot on Glorioso rather than Assumption. Glorioso was at that time covered with fine forest whereas Assumption supported mere scrub. This is important because Abbott casually mentioned that his booby nested on Assumption but he gave no evidence – just the throwaway comment on the strength of which Assumption has ever since been cited as the only known breeding place in recent times, other than Christmas Island. Nobody ever found Abbott's nesting on Assumption, whose scrub, as I have seen for myself, is absolutely unsuitable if Christmas Island is any guide. Did it or didn't it? We will never know.

So, with pent-up excitement, we watched Christmas Island come into focus and its seabirds materialise – that momentous first glimpse of an island's wildlife. Christmas Island sent out its special frigatebirds (Andrew's frigate), red-tailed tropicbirds with elongated central tail feathers that droop gracefully in flight, even the golden bosun, one of the loveliest seabirds in the world, especially when flitting through the sunlit foliage and black shadows of the jungle trees in which it nests. Two of the boobies – the red-footed and brown – were immediately conspicuous, but no Abbott's. In a way, I was glad. I wanted to meet it when we were out in the jungle – a crazy habitat for a large seabird. Abbott's is rarely seen amongst the common crowd, indeed it is hardly ever seen at sea at all. The moment when I spotted one in the Java Straits, some years later, was a highspot, especially because it was just where I had predicted.

Amongst the people on the jetty there was a stocky chap with a thatch of greying hair, like an old badger, and his left arm in a sling, irreparably shattered by a bulldozed tree. It was David Powell, later to become our best friend on the island and an indefatigable worker on Abbott's booby, but at that time a surveyor for the mining company. It was as well that neither of us had any inkling of the battles that lay ahead. Even as I write, my table is full of Christmas Island – maps, reports, submissions, correspondence and data.

Our house on the island, like our hut on the Bass, was

wooden, but that's the only thing they shared. Here we lived in a cavernous old mansion so big that we could play tennis in the living room. It had been the General Manager's house, the finest on the island, graciously furnished and hosting well-staffed dinner-parties. The guests had sipped their drinks on the verandah looking out over the roofs of the Malay campong to the waters of Flying Fish Cove, where it all began. Behind, on the forested face, the looming bulk of the phosphate drier fed the dust into the bulk of the huge pipe that led directly to the loading bay and the ships' holds. And that was why we were living there. Dust, all pervading, had driven the manager down into the settlement. The house was given over to the fat geckos, the moths and ants, and the garden to feral cats of astonishing ferocity – but to us, impecunious as ever, this free accommodation was a terrific boon. From our eyrie on the verandah we launched more than one Abbott's booby which had been grounded during tree felling operations and would have starved to death where it sat. We exulted as they flew steadily out over the sparkling cove. They were marvellous birds to handle – so tame and 'dignified'. A gannet goes berserk when handled and, the moment it is released, thrashes frantically away regardless of knocks, but Abbott's sit quietly, take their time, preen a bit, work their wings, and then, after careful preliminary inspection, necks elongated (a very necessary precaution when taking off from a jungle tree) they launch deliberately into a long-winged, oaring flight.

Abbott's booby has a majestic voice. A seventeenth-century French chronicler, referring to an unnamed booby on Mauritius that I think was certainly Abbott's, described it as a bull-like bellow. Whereas the gannet's voice is undeniably raucous, Abbott's is sonorous, like Paul Robeson's. The first time I really heard it we were sitting at the base of a noble, buttressed tree with the sun filtering through the translucent foliage and illuminating the tree ferns which glowed as though lit from within. The intimate jungle noises were all new to us – the rustle of a sidling crab, the sleepy 'roo-roo' of the fat imperial pigeon, the hum of insects, the caterwauling of the flying foxes or fruit bats. All this was swept aside in a breath-catching onslaught of sound, a pair of Abbott's, reunited at the nest and calling, shouting is a better word, till the jungle echoed. Nobody could have believed it was merely

209

Female Abbott's booby with chick. The white eyelid
is often drawn across the eye during display.

a couple of boobies; it was so deep, so expressive, eventually
subsiding into throaty grunts and queer, glottal clicks. All
the while, they were displaying with slow, powerful beating
movements of their great wings, reaching forwards and
downwards towards each other. They do not stand breast to
breast, fencing with their bills as gannets do. I believe that to
do so in the tree-tops would be too dangerous; better to stay
well out of range, avoiding physical contact as much as
possible. A fallen Abbott's is a sad sight. It simply resigns
itself and will quietly starve to death on the jungle floor.

Every facet of behaviour that diminishes this risk can be expected to increase the chances of survival in this curiously inappropriate breeding place.

Their appearance is as distinguished as their voice. From the very first glimpse as they fly high and steadily over the island, with their rakish, Concorde-like gizz, one senses that they are out of the ordinary – aristocrats. As a family, boobies hardly qualify for such encomium. They are quarrelsome, garish and clown-like. But Abbott's is different. It really *is* different in that it is highly aberrant. It has a distinctive profile, unusual habitat, unique ecology and very atypical behaviour. But these latter characteristics we had yet to discover. At this stage we were still thoroughly frustrated, chasing the sounds, gazing into the canopy and seeing nothing.

In my experience, the crucial time during fieldwork is in the early stages when one is setting up the study. All too soon, routine takes over and it becomes easier and easier to look without really seeing. Even with an unknown bird like Abbott's booby in which anything one can discover about its numbers, ecology and behaviour is going to be new and exciting, much depends on the questions asked, which in turn rests upon the set-up. If, early on, we had decided that the problem of getting to nests for detailed observation was too intractable the whole thrust of our work would have been different. Reaching nests certainly was a problem and we were not in the least surprised that nobody had tried. The rain forest on the island's plateau was a rich mixture of trees, all competing for sunlight. Consequently their boles were smooth for anything up to forty feet or more and since Abbott's booby often places its bulky nest well out on laterals, with air space all around, they were, to put it mildly, not easily reached. Since our study everything from mobile cranes to pylon hides have been mooted, but nobody has yet duplicated our tree-top work done with string and sealing wax.

On the scrounge, we rattled past the Chinese graveyard, with its little dishes of food set out for the ancestors, and on to the rubbish tip where rusty drums, old machinery and domestic rubbish mouldered in a riot of introduced weeds. Imagine a rubbish dump on the limestone terrace of one of the most beautiful islands on earth, with Andrew's frigates, unknown anywhere else in the world, nesting a stone's throw away. We were looking for packing cases, old crates or anything wooden. We had found three Abbott's booby nests in promising positions. One sat quite low in a banyan tree growing on top of a cliff, which gave it commanding height above the boulders littering the base. By hacking a way through thorns and creepers to the cliff, climbing a cleft and negotiating a steep litter of boulders choked with brush, we emerged onto a precipitous ridge hidden beneath giant ferns and snared in the roots and drapery of the banyan tree. A deep crevice obscured by ferns separated this booby ridge from the broken ground beyond. Ahead and slightly above us, through the tracery of leaves, a large dark patch stood out about eighty feet above the ground, viewed through a wild jumble of chasms and roots. Here, a trifling distance from the railroad, we were in the most forbidding sort of Christmas Island, a wicked alliance of razor rocks, and treacherous, long dead, fallen trees matted with an obscuring but yielding green mantle. Later, I had to lug a mountain of gear amongst this stuff and it was enough to try the patience of a saint. Soon there was a positive trail marking our route up the cliff and we knew every hand-hold.

The first time, we were not even sure that we had managed to emerge near the nest. A few yards in this stuff seems endless. But just before dusk the booby arrived, we climbed urgently to our new vantage point and, sure enough, almost hidden by the leaves but huge and near, above all near, was Abbott's booby, preening. It sidled onto the nest and began desultory building. All we had to do now was to clear the view, for the nest, far out on a lateral, was obscured by foliage and enormous tree-ferns clinging to the trunk. Behind us, the rocks were only a few feet down; in front and directly below yawned a fearsome drop onto limestone fangs. The booby tree leaned slightly outwards from the rock face but the roots penetrated the cliff so massively that there was no fear of the tree parting from the rock. By hacking above my head I cleared a few ferns, which smothered me in tickly dirt, and then, rung by rung, I nailed five well-separated steps to the trunk. Space billowed on all sides but one, and I felt the same awful clutch as when I once scaled the outside of a 212-foot factory chimney on a steeplejack's ladder – it seemed to be falling backwards, away from the chimney, which itself began to tilt, until my stomach tightened into a concrete knot. Eventually we commanded a good view of the nest from a crotch and, for the moment, this was enough, though I knew that one day I would have to venture out to the nest itself. From our perch we could see right over the jungle, to the sea far below. The incoming boobies would fly straight towards us, for that was the open side. It was dusk before we left our first Abbott's on his nest, hunched and motionless, facing the ocean. We could hear music from the settlement and, nearer, a train's horn blared deafeningly. Whether through conditioning or innate fearlessness, the Abbott's paid no heed at all. A few days afterwards, whilst we were working in the tree, improving access, the female came in. To my utter astonishment she apparently went peacefully to sleep whilst I hacked away at branches below her. Actually, she was drawing a conspicuous white lid across her luminous brown eye, but on my side only. The other eye was wide open. Far from sleeping she was shutting out the unpleasant sight of me with one eye and taking in more normal cues with the other. Later, we saw this 'false sleeping' in several social encounters. But she really was extraordinarily tame and on many occasions the pair relaxed, preened and built their nest, totally indifferent to us. We held no more interest or threat than a couple of monkeys. An hour after her arrival, the male flew in and, right under our delighted noses the pair enacted a marvellous and protracted greeting ceremony. The head feathers were used with great expressiveness. They can be sleeked back to give a long, sloping forehead or ruffed to enlarge and round-out the head to almost owl-like proportions. They faced each other and displayed slowly and powerfully, calling in that gloriously resonant voice. From the beginning the female's posture was inclined more markedly forwards and this became even more pronounced, the tail eventually beginning to lift and fan. She was soliciting and soon the male mounted, laying his bill alongside her head without gripping it whilst she grunted throatily. Both of them fluffed out their head feathers to the limit. This was the first close-range meeting ceremony and mating that we had seen and one of those special moments. Comparing all the social behaviour of the different sulids is like painting a complicated picture, which eventually provides a glimpse of evolution at work. But comparative behaviour is nowadays largely ignored by ethologists.

None of our other nests were as good as this one. Day after day I perched in the Abbott's tree waiting for the off-duty bird to sail in on long, rakish wings and take over the nest. These occasions were the most exciting I have ever known. There was additional spice in being the first to come to grips with this marvellous seabird and to be doing it on my own, my own boss, making my own films and completely free from the responsibilities of team work. And to be doing it on so little money pleased me – not because I had the funds and wanted to keep them, but because by doing it this way I was making it possible when otherwise it would have been out of the question. This may all sound egocentric and indeed I cannot guarantee that I could not have raised money for two or three other people to join in and make a proper team study. But I preferred it the way it was for good reasons as well as idiosyncratic ones, chiefly for simplicity and the chance to follow my own bent and methods. Nor was the avoidance of friction, always a real possibility when working with others under these conditions, a negligible gain. Finally, our good relations with the islanders and the mining company rested entirely in our own hands. As it had been on the Bass and in the Galapagos, so it was on Christmas Island.

At first, though, we were depressed about our chances. We had been uncommonly lucky both on the Bass and in the Galapagos to find our birds on the very doorstep and as accessible as barnyard fowl. Imagine being able to walk up to a tree-nesting seabird and simply pick it up. Yet that is exactly what we did with the red-footed boobies and frigates on Tower Island. But on Christmas Island the red-feet and frigates nested on the terrace trees between the sea cliff and the inland one. From the top of the latter we looked way down onto a billowing green canopy dotted with totally inaccessible red-footed boobies but once we got down there they were equally unreachable from the ground. Eventually we located a small group halfway down the cliff, in trees precariously lodged there. The Christmas Island frigates were just as bad. They nested in dead trees, dangerous to climb – all very discouraging. We were further dampened by the tail-end rains of the monsoon period; six inches can fall in a day. I am not a jungle-lover and the gloom and drip of the sodden rain-forest did as little to lift our spirits as did the

A female Andrew's frigate, one of the world's rare seabirds found only on Christmas Island.

infernal mosquitoes. A hundred large, itchy bumps on part of one arm is ninety-five too many.

This was our tuning-in period. Our Commissioners' jeep, without which we would have been stranded, visited many of our working places almost daily, along miles of rutted and overgrown track far from the settlement. We began to use this jeep in earnest, pushing deep into the jungle along tracks which were about as visible as the middle of a rhododendron shrubbery. The twigs thrust their wet and whippy fingers into our faces, soaking us and depositing thousands of nasty sharp little black seeds down our necks and up our shorts. Insects came in too. A large green mantid advanced along the roof turning its predatory head from side to side. A grasshopper with enormous antennae, which constantly tripped it up, became bogged down by them in the film of water and hung there, kicking frantically. The air crackled and pulsed with cicadas, several species singing simultaneously and no two alike. Underfoot, glistening red crabs sidled and pincered, as common as earthworms, whose niche they fill on the island. One approached head-on, a study in evil intent, two large, baleful eyes fixed unblinkingly in a heavy jowled red face, massive claws outspread. It stopped, turned a little sideways and was transformed. The eyes were painted on, sightless. The real things were on stalks. The large claws picked daintily at a morsel of leaf and transferred it delicately to the tiny mouth, precision-feeding the fragments into the palp-guarded orifice. They were everywhere, squashed by the million. We always tried to avoid them as they scuttled in front of the jeep, by turn menacing and prayerful, folding their pincers and waiting, motionless. They loved bananas and papayas but were not above feeding on their squashed brothers. Every year they migrate to the seashore to spawn, the males lying in wait near the water for the berried females. Nothing turns them aside at this time, nor on the return journey to the gloom of the jungle.

Christmas Island's beaches, until a few decades ago untouched for twenty million years, now bore man's ugly imprints. Lily Beach was littered with coke and beer cans and booby heads. The drink didn't occur naturally but the roast did. A mere eight years before our visit red-footed boobies used to nest in low scrub near Steep Point; there were none now. Brown boobies still nested near the beach, but their heads, too, frequently mingled with the beer cans and numbers were dropping sharply. The attitude of the Malays to wildlife was carelessly exploitative. From the youngsters who slew ground thrushes for fun, to their parents who slaughtered a sizeable portion of the world's population of the endemic Andrew's frigate with bicycle chains on the end of long poles, one and all saw nothing in conservation and everything in killing. Of course they didn't need the food, but it saved a few dollars for clothes or transistors. On 19 April we stumbled on thirty dead Andrew's frigates, or their remains, in a small patch of terrace jungle. It was a mournful place. The heads were stuck humourously in the forks of shrubs and a few were very fresh. All except one were males with the scarlet sacs which they develop in the nuptial season. A slightly immature male blundered through the understorey, dragging a broken wing. As we approached, his spiky head-feathers stood on end and his dark eye threw back a glint of sun. Once, he cocked his head to con a wildly displaying male, high in the greenery and in full vigour. This rare and beautiful seabird, which takes many years – maybe even ten years – simply to grow to maturity, is, to the Malays, merely 'a black bird – good for food – we eat him'. We drove back with him on June's lap. She thought it was docile enough to have its beak freed, whereupon it latched firmly onto her left nipple. We took it along to Government House to show Charles Buffet, the Official Representative, the results of Malay hunting. His wife Wanda was tending an injured bosun bird which was making good progress, readily accepting strips of fish from her fingers. She put a bit on the floor, proud to demonstrate her charge, and the frigate snapped its long, hooked beak menacingly. The bosun ignored the warning and pottered within range. The frigate struck instantly, gripping the bosun across the forehead and eliciting a wild screech. It tore itself free and, still shrieking, hurled itself at Wanda's ankles in furious, redirected attack. There was wild confusion, the frigate attacking the bosun which in turn was savaging Wanda. That was the end of her concern for the frigate, which came home with us where, alas, it died. No more idling in thermals, no more soaring, pin-head size, above its green island, no screaming dives with the wind battering every taut primary. Poor fellow.

It is ironic that, since the clearing of large tracts of jungle, Andrew's frigate, rare, lovely and confined to Christmas Island, has begun to cause deaths among Abbott's booby, rare, lovely and also endemic! Frigates use thermals for soaring, on the look-out for incoming boobies which they can sometimes rob of their fish. Cleared ground provides rising air and also deadly 'holes' into which a harassed booby can 'fall'. The safety net of the jungle canopy is missing. David Powell has seen Andrew's frigates force adult boobies down and has found several juveniles which almost certainly were similar victims.

A spry little adult male red-footed booby that we dumped into Wanda's lap became endearingly tame. It followed her everywhere, begging for food, and to my surprise it used the full, intense juvenile begging behaviour with feinting, head-bobbing, wing-flailing and a subdued begging call. All this

217

would have lain dormant, never used, for the rest of his life had he not been captured and fallen back into a dependent role. Yet, all that behavioural equipment would have been there; the nerve pathways, the precise, stereotyped muscle movements, the accompanying call – a complete pre-formed package. Of course, in many bird species which show court-ship feeding the females do use infantile begging behaviour to solicit food. But the red-foot does not have courtship feeding and this bird was not a female.

This small incident provided a glimpse of the sort of be-havioural material that is available to be used in social inter-actions, eventually, perhaps, to evolve into a 'new' display, just as a physical structure, used in a different context than before, changes into a 'new' implement.

The red-foot eagerly snatched strips of fish from a dish of water but obstinately refused to fly. After several attempts to get him air-borne, producing only a semi-paralysed effort and a belly-flop, he suddenly caught the wind and with a 'click' that could be heard a mile off, his reflexes took over and he swept upwards like an arrow. Two frigates immedi-ately latched onto him (did they know he'd just been fed?) but he quickly left them and soared out of sight beyond the cliff. Getting a bird to fly before it 'wants' to can be surpris-ingly difficult and later was to cause us some trouble when we tried to return fallen but free-flying young Abbott's boobies to their nests. The grounded birds understandably made no attempt to fly – they simply couldn't have gained enough height to get into the canopy. But even when we placed them on a suitable take-off point they took a long time to build up to the point at which they were ready to actually fly. In an earlier chapter I described how, in the young gannets on the Bass, flight-motivation gradually built-up until that invisible but very real threshold was reached. It is curiously difficult to understand that a perfectly undamaged bird, well able to fly, in the sense that its wings and muscles are in working order, may nonetheless simply glide to the ground if its nervous system doesn't *command* those muscles to work the wings. Yet it is so. Several years later, after much of the jungle had been cleared, one such youngster made a bad landing in its tree, which was in a partly cleared area and therefore lacking its safety net of surrounding canopy, and fell to the ground. We took it to a high bank nearby, placed

218

it on a crosspiece tied to a long sapling, climbed onto the top of a truck-car and, with all that height, launched it into the air. It simply opened its wings and glided down, like a paper aeroplane. We tried again, and again, with the same result. That bird became a long-term resident at David's place, MQ5 as he called it. With the next casualty I tried a bit of applied ethology. We put the bird on a high perch as near to the lowest bit of canopy that could be of use to it, and left it. That was in the afternoon. In the morning it was still there, but showing signs of rising flight motivation. Two hours later, of its own accord, it flew.

MQ5 was a marvellous place. David's garden extended to the low, undercut sea-cliff of blackened, razor-sharp limestone. Beneath the shade of some coconut palms he built low, twiggy platforms, and to these artificial nests the grounded young Abbott's were taken if they were weak or there was no way of knowing where they belonged. Several of his guests were newly-fledged youngsters which had foundered on or near their first flight and so had at least six months ahead of them during which they would normally have been dependent on their parents. Instead they became dependent on David, some of them for up to 194 days.

They were free to fly out to sea whenever they wanted and several of them used to fly out and then return, as they would have done to their proper nest, except that at MQ5 they had to land wherever they could make it, and be rescued. Even from the first, they were tame, although initially they had to be force-fed. People often ask me whether it would be possible to introduce Abbott's booby to a new island, but the difficulties would be great. First, one would have to hand-rear many scores of youngsters for several years in order for there to be any chance of enough surviving to breeding age. Each youngster would require some nine months of hand-feeding. Even then there would be no guarantee that they would return to their foster-island rather than the one on which they were born, although if taken before their first flight, there might be a fair chance. Introduction, or re-introduction, is just beginning to work with the introduced sea-eagles of Rhum, in Scotland, but one must remember that these birds are returning to an area in which they used to nest, where the terrain is known to be suitable.

Although brown boobies are common on Christmas Island, as in the tropics generally, they do not nest in the Galapagos and we were not familiar with them, so they, too, figured on our 'wanted' list. We set off for the brown boobies of Lily Beach and Steep Point with generous loads. June carried three gallons of water and a hefty bag whilst I had a rucksack, tent, ciné camera and tripod. It was little more than a mile but a sizzling one in the late morning. The spiders had strung masses of webs across the trail and we collected them all on our sticky faces. The coral-shingle beach, blindingly white, emerged suddenly after a sharp turn along a rock face; one moment you were among the trees and the next, out onto the beach. Off with our clothes and in for a cooling bathe.

Sleeping on the beach, even with a layer of leaves, was distinctly uncomfortable. The rats were busy all night and large hermit crabs scraped and bumped along with their shell houses on their backs. Robber crabs and red crabs were out in force and altogether I was happy to climb onto my feet at first light, but it was a grey dawn with nothing doing and nothing worth filming. Steep Point is a high jagged ridge difficult to negotiate and obscured with vegetation. A blinding rainstorm caught us on the way down and it was a disgruntled and jaded pair that crawled out of Lily Beach at dusk and headed for the settlement.

We returned a few days later. Where we had camped and left without a speck of litter there was the debris of a Malay picnic – cans, bottles, wrappers and the fresh heads of two brown boobies and five noddy terns. At that time, some two to four hundred seabirds were killed every week (that was the conservative estimate of two very experienced islanders who knew the Malays better than anybody else before or since). It was simply impossible to police the island for bird-killing even if the Malay police had been inclined to bother, which they were not. Twenty years earlier, when the imperial pigeons were abundant, they sold for 1p each and up to seventy-five were shot in one tree. However, no Christmas Island bird has been seriously threatened by hunting and, on the contrary, one or two introductions have flourished. The Nankeen kestrel, a delightful little falcon, had arrived only twenty-six years before and was now abundant. In fact I felt no real concern for the pigeon because it had the capacity for explosive increase whereas slow-reproducing

The brown booby often lays two eggs but, like the masked, ends up with one chick because of sibling murder.

and long-lived seabirds do not. Rare or common, indigenous or introduced, it was all the same to the Chinese. Even the kestrels went into the pot and fruit-bats were gathered by the hundred. The red-crab migrations provided unlimited poultry food for weeks on end; the tiny crabs swarming up into the jungle from the sea-spawning filled a bucket in no time if a plank was placed as a ramp. Fishermen stripped the females of their berries, even though this killed them, as bait for a species of fish that came inshore especially to feed on the crab larvae.

There was another splendid little group of brown boobies down in the south-west, near an area wistfully called 'The Dales', though anything less like the Yorkshire Dales could hardly be imagined. On one occasion Wanda Buffet accompanied us to this spot, reached by trekking two or three miles through the jungle on a trail which, at times, was easily missable. It was blazing hot and by the time we had lugged the usual paraphernalia down to the colony and worked all afternoon in the open we were pretty jaded and looking forward to the iced drinks which we had left in the jeep. We gave ourselves more than an hour of daylight to get back and set off, a bit leaden-footed, out of the cove. Quite soon the faint trail petered out and stopped. Even the odd blazed tree no longer turned up to keep us pointing in the right direction, so we retraced our steps and started again, with exactly the same result. By then it was gloomy dusk and we were thirsty and fed-up. Soon it was dark and we were beginning to trip over the roots and stumble into muddy crab-holes. The jungle seemed a miserable place to have to spend the night so we decided to return to the cove, although it was by now pitch black. Luckily I carried matches and the blank pages from my field notebook provided short-lived torches. Eventually we could hear the sullen boom of the surf battering the deeply undercut sea-cliff. There was only one tiny access to the cove, through otherwise impenetrable thickets of saw-edged pandanus and by sheer luck we found it and were back at the point we had left a couple of hours before. We lit a fire and found a rusty old beer can in which to boil some of the water – half mud – which we had scooped up from a blue crab-hole. Filtered through a handkerchief it yielded a turbid but drinkable liquid. There was no level ground to lie on so we hunched away the night hours and set off again at dawn. This time we were successful – a fallen tree barring the real trail had been our undoing – but, alas, the search was already under way and my reputation was as muddy as the crab-water. Charles Buffet never forgave me!

About this time, on account of gremlins, I almost threw

my camera into the Indian Ocean. May 27 was typical. We set off in the morning to our best Abbott's booby tree for June to film some linking sequences, such as me approaching the tree and climbing it, since you can't suddenly show Abbott's booby on its nest with no hint about how you got there. The treacherous pinnacles, covered with creepers and ferns, gave no footing either for the tripod or the cameraman but we got it precariously balanced and the camera set up and off I went, authentically laden and more than authentically tripping and slithering. It would look good on film. 'I don't think I'm focused' (June), so I return and set off once again. Damn, this time the sun's gone in. That sun thwarted me at the critical moment so often that I became positively paranoid. Like extra-sensory perception it went way over the statistical odds. To reach the tree I carted the gear up a face so steep that with a slight forward lean I could, and did, brush the sweat out of my eyes with my knee, which was just as well since both hands were full. The showpiece was to be me climbing the tree, but this time the film hadn't been winding-on. I never *was* filmed climbing that tree. After that fiasco I spent four consecutive afternoons up it and on each occasion nothing happened until I left, whereupon the absent bird came in and the greeting-ceremony that I'd wanted, occurred. And it wasn't as if I could easily saunter along for another try. Fixing three tripod legs on three branches, seventy feet up, without a platform, is a bit like doing a Rubik cube standing on your head. On the fifth occasion, just as it became too dark to film, and I had laboriously dismantled everything, the male came in and fed the chick, which dropped a big fish and put on a great show picking it off the floor of the nest and choking it down. Gremlins.

It was now mid-June, about halfway through our stay, and we were beginning to have ideas about why Abbott's booby nested where it did. They seemed to favour areas with conspicuous topography. A steep but uniform slope with no breaks in the canopy didn't attract them even if it rose to more than five hundred feet, below which they rather rarely nested. But where a spur, or a sharp rise produced marked irregularities, Abbott's were more likely to be found. Tom's Ridge, Murray Hill, Wharton Hill and the centre of the island, which presented a ridge from the south, all possessed this feature and were prime Abbott's areas. Emergent trees

223

offered easier entrance, especially on the north-west, since Abbott's could land into the south-east trade which blew for much of the year. This suited this long-winged seabird which cannot float like a gull, or even a frigate. Often, the jungle changed dramatically as one descended from the plateau. Within a few hundred feet, covering only a slight drop, the average height of the canopy could change significantly. Abbott's seemed to prefer a canopy whose average height was more than forty or fifty feet although often nesting far above or quite a bit below that height in individual trees. Large tracts of jungle between five hundred feet and sea-level have a relatively low canopy height and may be neglected by Abbott's for this reason.

I had the idea, too, that a discontinuous and visually rich area would offer valuable landmarks to a young booby homing-in on the nest after its first flight out to sea. Imagine the problems facing such a youngster, flying in for the first time over a vast, uniform canopy. David didn't think much of this idea but I knew that young boobies in general imprint on their immediate surroundings long before they ever leave the area. They can find their way back from any direction. I realise that seabirds have a fantastic ability to locate their own spot amidst a dense mass of neighbours, often without obvious landmarks, but this ability comes only after hundreds of prior visits. For the young Abbott's, the first return may be critical. It spends many hours, outposted high in the canopy, before taking its first flight and it is my belief that it fixes the landmarks at this time.

The distinctly 'clumped' nature of Abbott's distribution, though falling far short of any other booby in this regard, was not due solely to the presence of suitable jungle; their undoubted gregariousness played a part, too. Darling's interest in the social side of breeding behaviour has been mentioned in these pages more than once and it still remains one of the least studied and understood aspects of seabird life. What Abbott's booby gets out of it remains to be shown but an improved chance of finding a mate could be one advantage.

One evening, David mentioned, over a bottle of beer, that Abbott's often flew in from the north-west but apparently not from other directions. This immediately caught my interest and we began to spend a lot of time at the far end of Tom's Ridge, scanning the late-afternoon sky for those distinctive silhouettes (they came in mostly during the last two hours of daylight). Sure enough, in they drifted, often speck-high, scarcely visible to the unaided eye. To our astonishment and jubilation we began to clock substantial numbers, far more than we expected. On one day we counted over 2,000 incomers. Several years afterwards David began similar counts and his record, at 3,259 in a single day, was much higher. They were not all breeding birds of course. Some would be adults taking a year off, some young birds, may be two, three, four or even more years old whilst others may have been dependent but free-flying juveniles.

My diary for 3 August 1967 records that as a result of Tom's Ridge counts I guessed there would be not less than 1,200 adult pairs and several hundreds of juveniles on the island. Fourteen years later, after an island-wide search that took David and Kim Chey, his Chinese assistant, thousands of man-hours of fieldwork, they came up with a figure of 1,136 pairs of Abbott's involved in breeding in the year 1982 (a single breeding effort, if successful, takes *two* seasons, so that in any one year there are some pairs with eggs or new young and some with old young from the previous year, so that 'engaged in breeding' means tending eggs, or young of any age).

I suspected that some of these young pre-breeders mentioned above were responsible for the puzzling 'threesomes' that we frequently saw. Almost always the extra bird was a grey-bill, apparently a male. What more likely than that the juvenile from that particular nest should, on its return to the island, home straight onto the tree in which it had sat for some 5,000 hours and to which it had returned hundreds of times during its dependent phase? If its parents were still using it, which is quite likely since the same pair often use the same site for several years in succession, a threesome would result. It wouldn't last, because the territorial display of the owning pair would drive the intruder away, but the frequency with which threesomes turned up argued for some such explanation.

Long before we dreamt of counting the island's population we concentrated on our precious study nests, which eventually held their large, white eggs (a tremendous thrill, this) and then their growing chicks. By now we were climbing our booby trees every day. Our oldest chick, alas, fell fifty feet

225

whilst defecating tidily but unsteadily over the nest rim, doubtless a hygienic necessity but one which had often given us palpitations. We had designs on the adults as well, although it meant catching and weighing them high in the tree rather than struggling all the way down and back again. From the crotch below the nest the male's beak was just visible as a blue dagger-point. June wedged herself against a limb and I hoisted myself the last few feet. I needed both hands to deal with the adult and had to make do with an arm hooked round a branch for my own hold. He made no attempt to fly – far too dangerous – and (mercifully, for they have large, saw-edged and powerful bills) no attempt to bite. Despite his dignity, he had to be transferred to the weighing bag, with June's invaluable help from her wind-swayed perch, sixty feet up. Understandably, he vomited a bolus of warm and slimy flying fish which slid neatly down inside her blouse. Even so, when we put him back he settled stolidly onto his nest; once again we relied on Abbott's strong inhibition against any hasty or panicky movement in the canopy, a trait which served them as well as it did us.

It was not long before we were putting in a fourteen-hour day, much of it spent lugging equipment around over awkward ground in the humid heat. As usual, my eyes were greedier than my stomach. Andrew's frigate received scant attention although I did rig up precarious access to a group of fourteen nests in a part-dead tree. The locals called it a cabbage tree because its branches snapped so easily. I had to reach a limb about twenty feet high in order to throw a rope over a fork a further fifteen feet above. When I did get it over and fixed the rope ladder, the route passed rather too close to a male's champing mandibles, which actually parted my hair as I crawled past him. I soon began to hate that tree.

Weighing frigates, a solo effort, was the worst job. The rungs climbed up a backward-leaning limb and the nests were only just within my most extreme reach. The chicks were by now hefty lumps of some 2 kg and they clung grimly to the excreta-cemented twigs which served as a nest, snapping furiously with their long, hooked beaks. Then they tried to vomit their last meal, a calamity which had to be prevented at all costs – another full-time job for two hands. Worst of all, they had to be coaxed into the weighing bag to be strung up on the spring balance. The bag blew around in the wind, infuriatingly close-mouthed and offering a perfect target for those sharp little claws. As soon as one desperate foot was prised loose the other grabbed hold and all the time one hand was needed to keep the bird's mandibles closed. I never earned anything with more sweat and bad temper than the growth curves of those four frigates. One female parent left its nest and attacked me from the air. Her first swoop dislodged my floppy hat, which floated down to a branch twenty feet below. The second attack struck my scalp. Third time round I grabbed her long beak, snatching her neatly from the air in full flight whereupon she continued to beat her great vanes furiously, nearly taking me with her. At that moment I had my camera in the other hand, which left me in a bit of a fix until I got the camera strap between my teeth. In retrospect the risks seem to have been quite ludicrous but at the time seemed natural enough.

As we wandered around looking for Abbott's we came across several young golden bosuns on the jungle floor; at that age they are lovely silver birds closely barred with black. The grounded ones were the failures, the ones from which the crazy nesting place had, not surprisingly, demanded too much. As I have alraedy mentioned, the golden bosun leaves the open sea and sky to risk its life looking for a nesting hole in a jungle tree. When the fledgling is ready to leave it has to find its way down to the sea. Its options are stark indeed. It can't practise flying before making the journey since the risk of a mistaken landing on the jungle floor, which could be fatal, would be too great. It can't keep on returning to the nesting hole whilst gaining strength and experience for it would be suicidal to attempt to find its way back maybe more than a mile inland and beneath the dense canopy, and repeatedly at that. Any fat reserve that it may carry would make it dangerously heavier during that crucial first attempt when wing loading must obviously be critical. Yet without fat its chances of making the transition to independence at sea are inevitably reduced. Its parents do not accompany it and so it cannot look to them for help. Despite all this, the simple fact that the golden bosun bird has maintained its numbers on Christmas Island for thousands if not millions of years shows that evolution's answer has worked, but the Lord only knows how.

However, it was the problems of Abbott's booby, rather

227

than the golden bosun, that chiefly occupied us. Coming to grips with their breeding regime was reminiscent of our headaches with the frigates of the Galapagos. It may seem as though we arrived on the island, found the boobies about to begin nesting and simply followed through from there. But in fact we were thrown off-course immediately because we happened to see a juvenile begging and being fed, only we didn't know it was a juvenile. We didn't know – nobody knew – that the juvenile Abbott's, whilst still dependent on its parents, is virtually indistinguishable from its father. No other juvenile booby resembles the adult male. So when we saw the ritualised begging we thought we were looking at a displaying male. That pardonable mistake was soon cleared up when the presumed male was seen to be fed by a pink-billed bird which we knew to be a female. No booby shows courtship feeding and, in birds that do, it is the male that feeds the female, not vice-versa. This helps her to meet the energy-cost of producing eggs. So when we realised that she was feeding a juvenile we cursed, and jumped to the equally wrong conclusion that we had arrived just in time to see the curtain falling on the breeding season. Then we saw a male land near the top of an emergent branch and begin tugging at a leafy twig. Clearly, therefore, some birds were just beginning to build whilst others were engaged with free-flying offspring. From that moment, everything began to fall into place. By good luck, we had timed things perfectly. Some pairs were beginning a breeding cycle whilst others were finishing one. We got them both.

The free-flying young had in fact hatched some time the previous year but, like the frigates, were still dependent on their parents. The one we could see best used to outpost itself high in a tree, a white shape in the foliage, difficult to distinguish from a patch of sky seen through the leaves. As soon as a parent arrived the youngster hopped and sidled down the branches until it came to one particular crotch. There, and nowhere else, it was fed. Of the nest which had been there not a trace remained. The monsoons had washed and blown it all away but the nest site remained the only place at which the adults would deliver food. Before that happened the youngster had to beg, and beg, and beg again, with a bobbing, swaying motion of the head accompanied by a harsh, repetitive call of three or four short syllables followed by a

228

grating croak – 'aa-aa-aa-aaah'. This became the most familiar sound of Christmas Island. During the last two or three hours of daylight we used to walk the jungle survey lines listening for begging youngsters. Their calls were positive evidence of a breeding pair, even when we couldn't see a thing. Although we couldn't know it, the period immediately following our arrival on the island was the one in which food was scarcest and most youngsters died.

The months slipped by but still the free-flying youngsters from the year brefore hung on and still their parents kept on feeding them. By August, we could scarcely believe that juveniles which had been fully on the wing in February, and had presumably hatched sometime around July of the previous year, were *still* not ready to cut the apron strings. Then, in August, they abruptly disappeared. One day they were begging as usual; the next day they were absent and we never saw them again. They had been dependent on their parents for more than a year, which meant that Abbott's booby could not produce a successful offspring every year but only once in two years. This explained, of course, how it came about that we had two sets of adults on the island – one finishing the breeding effort of last year and the other starting their breeding effort this year. It was obvious once we had the necessary bits of information – it always is. But there was still one thing which greatly puzzled me. According to all accounts, Abbott's booby quitted the island during 'winter', the monsoon period, reappearing in March or April. But if some of them had large offspring still requiring food, how could the adults quit the island? David Powell had told us that he would be working in the jungle, never seeing or hearing a booby until, suddenly, they were back. The answer was simple. During the monsoons the adults with dependent offspring visited the island to feed them, but their visits were sometimes few and far between and they didn't remain as long at the nest as during the 'summer'. Consequently the pair met up less frequently and there was less chance of hearing their marvellous greeting calls. What we did not know was that most of the youngsters failed to last through this stringent testing period and simply died of starvation or were blown out of their tree by high winds. Only later work by David revealed this.

You may consider that Abbott's booby is a queer bird.

When I described it as 'aberrant' I understated the case. One could easily call it crazy, although that would be quite the wrong adjective. Yet does it not seem perverse to nest high in jungle trees and even more idiosyncratic to arrange things so that its chick has to squat on a bare branch throughout months of torrential rain and gusting winds, including the odd cyclone? And why do the adults appear to take their responsibilities so casually, allowing sometimes more than three-quarters of their offspring to starve? But all these strictures imply that Abbott's has a choice and could behave differently, which is untrue. Of course I don't know why it took to trees, but it did so a long, long time ago. Fossil boobies go back something like sixty million years and there were certainly more species and probably far more individuals, at one time in booby history. Most of them were ground-breeders and so the trees offered a usable niche, far from perfect, but at least out of the way of other species. The red-footed booby took a similar step and so did several other seabirds, including noddies, fairy terns, frigatebirds and a bosun bird – all of them, except the fairy tern, nesting in Christmas Island's trees to this day. This is where, it seems to me, evolution displays its devastating objectivity. If all the many adaptations which these various species display and which enable them at least to make a go of their unusual breeding places, were the result of some non-evolutionary act, like supernatural creation, why are they so imperfect? Abbott's booby is 'adapted' but it fails dramatically often.

By taking to tree-nesting, Abbott's booby 'condemned' its chick to sit on a bare branch throughout the monsoons, if its nest disintegrated. But, one may ask, why could it not avoid this unpleasant necessity by bringing its chick to the point of independence *before* the monsoons break? After all, the brown and red-footed boobies on Christmas Island manage to do so. Whilst the Abbott chick is fighting its battle, the young red-foot and brown booby is wandering at sea, foot-loose and fancy free, fishing whenever it wants and away from the dangers of land. The answer is that Abbott's grow too slowly to achieve this. The egg requires far longer incubation than that of any other booby and the chick continues this slow pace, taking six months before it can fly. The gannet achieves twice the growth in half the time! Having reached December or January before it can even fly, the young Ab-

230

bott's is between the devil and the deep blue sea. If it stays it may starve. If it goes, it is on its own in the monsoonal Indian Ocean, unpractised and inexpert. But why *does* it grow so slowly, the cause of all its troubles? Because it is fed relatively infrequently, on relatively small amounts. It doesn't receive the 3,000 g of oil at a sitting which the young albatrosses in the Galapagos did, not even, usually, the 500 or 600 g of mackerel which a lusty young gannet can tuck away. The fact is that the young Abbott's booby is uniquely adapted to grow slowly and to withstand long periods – maybe weeks – with little or no food during 'winter'. This is what it is good at and this is what enables its parents to take the time to forage a long way from the colony, especially during the monsoons (but don't forget that if only it could have grown faster in the summer it wouldn't have *needed* this fasting ability). So why has the Abbott's evolved its particular forag- ing method? One can ask the same question of any species and fill a library with attempted answers. It has to do with the evolution of diversity in living things. But, clearly, Abbott's has crossed the thin line which separates growth which is fast enough to achieve all that is necessary in one season, from that which just fails to achieve this. And even if it only *just* fails, that is enough to force it into its present course. There can be no question of abandoning a chick which is *just* too young to survive on its own. It is therefore fascinating to learn that, according to plate tectonics, Christmas Island has moved several hundreds of miles during the last million or more years. Maybe Abbott's has to travel further to reach its traditional feeding areas than it used to, with all the effects that would be entailed – fewer visits, slower growth and youngsters sitting out the monsoons having just missed the bus. Not an altogether convincing scenario but attempts to answer this sort of question rarely are.

An alternative scenario could envisage oceanographic changes such that the productive upwellings in the Abbott's feeding areas began, at some time in the evolutionary past, to warm up at certain times of year, thus making food harder to obtain, slowing growth and leading eventually to the exten- ded cycle that we now see.

To return to earth and to Christmas Island, another in- triguing thing about Abbott's booby is that, as mentioned, the juvenile can hardly be distinguished from the adult male whereas the juvenile gannet could hardly be more different. Now why on earth should that be so? Delving into this shows just how the jigsaw puzzle approach to comparative behaviour can work.

The black plumage of the juvenile gannet interested me from the very beginning because it seemed odd that whereas almost all plunge-diving seabirds are white below, very much including the adult gannet, the juvenile gannet is blackish. If white underparts act as hunting camouflage why should the young gannet lack them just at the most vulnerable period of its life? More young gannets starve during the first month or two of independence than at any other time. One might think that black plumage is better at absorbing heat from the sun and perhaps at keeping heat in the body but an experiment in New Zealand showed that it didn't do the latter, whilst the former seems of marginal benefit since the young gannet migrates straightaway to warm climes anyway. In any case, neither of these possible advantages seems likely to outweigh that of being less visible to fish and therefore more successful in plunge-diving. And *that* camouflage requires white plumage, not black.

A completely different tack was suggested by the adult male's habit of attacking the female each time they met. I described this in an earlier chapter. To watch the male of a well-married pair arrive at the nest and knock its mate clean off the pedestal with the sheer vigour of his attack was vivid proof that a single such incident, directed against the chick, would be the end of it. And since the male reacts thus to his long-standing mate, why shouldn't he do the same to his youngster if – and the 'if' is important – the youngster presents the same stimulus as the female? The female triggers attack because she presents a particular cue; she looks like a male and the male gannet is programmed to attack intruding males. This is all part of the gannets' rigid social system and she can cope by various behavioural means. If she is knocked off the ledge above a three-hundred foot drop, she can simply fall away into flight. The youngster cannot. One good way in which the youngster can minimise the danger of such an attack is to present, not the male-like appearance that invited aggression but quite the opposite, and this is more or less what it has done. Bear in mind that after the youngster has achieved the size and the stance of the adult and therefore (in cardboard cut-out) looks like one, it must survive scores of visits by the male. Colour is its only means of inhibiting attack. Appeasement behaviour would be no good because it would be too late. It is at the instant of landing that the male must 'decide' to attack or not. Any plumage feature of the young gannet which tipped the balance against attack must have considerable survival value. If it boils down to a trade-off between the advantage of hunting camouflage and the value of attack-inhibiting plumage, then that is the way it has to be. You can't have everything. In fact, such a trade-off would predict that, as soon as the youngster was safely away, it would turn white as quickly as possible on the all-important ventral surface (the one that camouflages the bird against a pale sky when seen by a fish). And this is exactly what happens. The underparts turn white within months, whereas the back stays black or partly black for years.

So far, so good. But what about Abbott's booby? First, the male Abbott's is amazingly unaggressive. By this I mean merely that it very rarely fights and usually even avoids physical contact with other individuals – even its mate, except during actual copulation. Its displays are usually conducted at a distance. Even that most provocative of intruders, an intruding male actually on the nest, does not provoke the owner to attack. How utterly different from the gannet. Now, if one construes the gannet's aggressive responses as triggered by external cues – that is, the sight of another gannet in the 'wrong' place – and the Abbott's responses as being *internally* controlled, the observations begin to make sense. Imagine a threshold which has to be crossed before an attack will occur. That threshold can be high or low, and can be influenced both by internal events and external ones. The gannet's threshold for attack is extremely low, so low that the most minor infringments of territory call forth strong aggression, so low that even its own chick, *off* the nest, can be torn apart; so low that a faithful partner of maybe ten years standing is regularly attacked. All these attacks, however, are coped with by appropriate appeasing behaviour and the ordered social life of the pair and colony is not disrupted. The Abbott's threshold is extremely high – so high that even intruder males call forth only weak aggression rather than all-out attack. There is, in Abbott's, a greater internal 'block'. Of course the external cues reacted to by the gannet have necessarily to work *via* their internal effects – they can do no other – but the point is that equivalent external cues do *not* have equal effects in the two species. The consequence of Abbott's internal block is that even if a young Abbott's booby does look like the adult male, it is not likely to elicit an attack. So, if there are benefits to so doing, it can afford to omit (or telescope) the juvenile plumage and adopt the adult plumage straight away. If this plumage helps the adult in fishing it is likely to help the inexperienced juvenile even more. Again, it is a trade-off.

Of course, this is a simplistic account of a much more intricate web of relationships and it isn't necessarily correct – it is merely a plausible scenario, perhaps interesting but to be disbelieved if there are serious shortcomings in relation to the evidence. But I repeat – such scenarios are not unscientific, they are a help to understanding complex behaviour. I cannot accept that they are merely unjustifiable speculation – as some mediocre 'hard-nosed' biologists with an undue reverence for facts alone, however stodgy, would aver. I say 'mediocre' deliberately because no outstanding biologist ducks the challenge of interpreting what he or she finds.

About a month after our arrival Lim Hoon, a Chinese bulldozer driver, pushed over a tree with two Abbott's boobies in it. This tiny incident encapsulates the sort of thing which has been happening on a gigantic scale, world-wide, this century. We want something and to get it we destroy whatever lies in our path. When first I watched a noble jungle tree uprooted by a bulldozer it left a deep impression. It wasn't simply the facts of the exercise – the metallic grinding and roaring of the huge machine as it thrust irresistibly against the living tree, the deep groans as the roots were torn out of the red earth, the rending crash as the branches tore through the surrounding foliage – but the symbolism that bothered me. Of course men have been felling trees ever since they had stone axes but for almost all of that time the earth seemed easily to absorb the punishment. Lately, and increasingly, it has been a different tale and we know it. The earth is now battered and the sight of a rare and lovely creature like Abbott's, clinging uncomprehendingly to its tree as the jungle is torn up around it, finally hurtling down in the avalanche of branches and ending up on the ground, muddy and trailing a broken wing,

234

just about wraps up all that is happening. Good old progress. I found the destruction of the jungle more traumatic than anything I had seen and that was a mere foretaste of what was to come.

Others reacted differently and the communication gap yawned as wide and indifferent as it always does. There was one mining-company surveyor, a portly young German with a pallid face, who had the drawing-board approach. If a new road had to be driven from A to B through virgin jungle, he tackled the job robustly – no keyhole surgery for him. Make it wide enough for ten Euclid trucks abreast, clear an extra hundred yards width for possible powerlines, add a few hundred acres for possible dumps, etc., and square the whole thing off. The massive, ruler-straight swathes of bare earth were soon transferred from the drawing board to the ground, contemptuously dismissed as 'bush'. Let's clear the 'bush'. Why not? We've got the machinery and the men and they'd only lie idle. Might as well use them. After all, we're here to mine, not to watch dickey-birds. But in 1967 most of the jungle clearing was still to come although it was already menacingly clear that a large part of the best phosphate lay under the jungle precisely where Abbott's boobies were thickest on the ground; perversity, perversity.

Alas, in those days I was just beginning. First I had to find out where the boobies were, and the island is large. Then it was necessary to demonstrate that the boobies and the phosphate did indeed overlap – at first the miners claimed there was no such problem. Next I had to show that Abbott's booby really was conservative in its distribution and was unable merely to transfer its affections to other parts of the island. I knew this simply from the way it obviously ignored huge stretches of jungle on the eastern side of the island and concentrated in the irregular and dissected terrain of the centre and west, but knowing it and showing it are two different things and the latter took time. However, it proved to be so. The next and most protracted phase took me well out of my depth because it involved the politicking and inter-departmental jockeying within Australia. After all, we were dealing with a multi-million dollar enterprise and a reasonably large island population. It would be tedious for the reader, and worse for me, even to summarise all that went on; I may try some day, but the upshot was that we did get a National Park established and, most vitally, a binding procedure which safeguarded at least most of Abbott's remaining areas. It required interest and action at the highest level, which meant Prince Philip and the Australian Prime Minister (Malcolm Fraser), but we did it. Whether the agreement will last remains to be seen.

The natural riches of Christmas Island are inexhaustible. Seabirds are but a part of the endemic bird life and birds a mere fraction of the island's animal life. The plant-life, including more than two hundred tree species, is largely uncharted and the staggeringly prolific and varied marine life is virtually untouched. The island has complex cave systems with vast supplies of underground water, and the surface geology would keep a professional busy for years. All this compounds the tragedy of its destruction, even though, blessedly, this is still only partial. Really, it ought to be an international heritage, along with other truly outstanding natural gems. The island that remains – which is still most of it – is rich, beautiful and biologically important. Even the mutilated areas could be restored and indeed, thanks to David Powell, some parts are being replanted. But there are far too many mined-out areas for the present scale of operations ever to cope. And, most gravely disturbing of all, there are proposals to inflict further extensive damage by more clearing, extending down even to the hitherto untouched terraces. Anybody who knew Christmas Island in the old days must wince at the scars, when they fly in nowadays. (I still can't believe there *is* a full-scale landing strip for commercial jets!) But I wince even harder at mining proposals which would destroy yet more prime habitat, to say nothing of the nesting areas of commoner species. Another suggestion was to clear a huge block of virgin jungle for farming so that there is some future for the island when phosphate runs out in the near future (maybe less than five years – the present-day economics are fluid and uncertain). This might sound reasonable, but it would be utter folly. Small experimental plots should be farmed for years first, before the irrevocable step of clearing this huge block is taken. The soil of tropical rain forest is notoriously fragile and subject to erosion and it would be more than foolish, it would be criminal, to discover that such a project was ultimately untenable, when the damage had been done. When mining

236

A mined-out part of Christmas Island. The bare limestone pinnacles had pockets of phosphate between them and were originally covered in rain forest.

does cease, the ensuing few years will be critical for the long-term future of the wildlife on Christmas Island.

Interestingly enough, the social changes which recent years have brought even to Christmas Island are only too familiar. A dramatic rise in material standards has coincided with a marked change in values and attitudes. Workers now come to get as much money as quickly as possible whereas formerly it was more for the security and the chance to put down roots. Nowadays few people stay very long. Expectations are high and money-rackets, formerly undreamt of, have been imported by a new breed of Mafia-like entrepreneurs. Resentments amongst workers now bubble over amidst threats of arson and violence which only ten years ago would have seemed like sheer fiction. The goose that lays the golden egg is threatened.

The truth is that Christmas Island is in deep crisis, right now. Mining has run into massive problems and, as I write, is losing millions of dollars a month. Enquiry after enquiry has looked at the problem and, after the company had almost decided to put up the shutters, a final enquiry decided that mining could be viable. Large-scale reorganisation ensued, but still the troubles continue. Unionisation and its insistence on paying the Asian labour-force wages equal to those of mainland Australians has undoubtedly been a potent factor, coupled with chronic over-manning stemming from the cheap labour days. Both union and management, although deeply divided, can at least join forces in blaming the troubles on conservation, with its proposals to limit environmental damage by protecting certain areas even though they contain some good phosphate. The politicians have grown weary of the incessant in-fighting, the endless enquiries, the media disclosures of idle workers (some may have done nothing except draw their pay for almost two years) and in May 1984 the minister issued an ultimatum. Either the island put its affairs in order and became viable or the mining closed down. At the moment it looks as though the loser will again be Christmas Island and its wildlife. There is pressure on the Australian Government to rescind the protection offered by recent legislation and allow mining in protected areas. If that is the price of peace – even a short-lived and illusory one – it may seem payable. Like the Galapagos, Christmas Island deserves full protection from *all* forms of exploitation, not just mining. Various government departments have a bureaucratic stake in this innocent speck of land and the next few years may well be vital in shaping its future. It has already paid its ransom price in millions of tons of phosphate and should now be spared further despoilation. The south-west corner is already an Australian National Park. The whole island should now join it. Long may its sunlit jungle reverberate to the resonant voice of Abbott's booby.

16

Back to the Bass

This observation, like all other observations of animal behaviour, however trivial they may seem to be, gives rise to the question upon which the scientific study of behaviour, or ethology, is based: why does the animal behave as it does?

Niko Tinbergen, *The Study of Instinct*

Never go back, they say. But I have never left the Bass. I go back each year to the place 'wherein there is nothing . . . that is not full of admiration and wonder; therein also is great store of soland geese . . . these foules do feed their young with the most delicet fish that they can come by.' It is true that the ancient chapel is no longer my home; its stones sightless where once they pried intimately into our lives in the hut. The February gales come and go without affecting me and I am no longer there to see my gannets return from their winter wanderings. But I have followed their fortunes nevertheless, not in the detail that used to be my bread and butter, but in many essentials. Some of my ringed birds are still there, more than twenty years on. And the growth of the gannetry has not gone unremarked. Indeed it has been spectacular and the vast snowfield on the north-west face, one of the finest seabird sights in Britain with more than 3,000 sites, now obliterates the track we trod countless times. If I placed my hide where it used to stand on a bare hillside it would now sit in the middle of thousands of birds. The gulls that plagued them have been displaced with a vengeance and the lush grass has long since disappeared, together with the soil, leaving the naked basalt which not even gannets can remove, though they do dislodge boulders, sometimes with drastic consequences.

I do not see gannets quite as I did at first. In a way, the freshness has gone. It is hard to look the thousandth time with the same riveted attention that marked those early sessions when everything was new and exciting. But to compensate I know much more about them and their relatives and can answer some questions that I couldn't even ask in the early days. So these pages are about insights, mainly into behaviour, that I owe to gannets. On reflection, perhaps I should not write 'insights', but merely 'ideas', for it may be that others find them anything but insightful. I shall return to the Bass and rough out the balance sheet. On the cost side has been the years spent on gannets and their relatives when I could have been playing golf, growing onions (I prefer onions to cabbages) or making money. On the profit side there have been the experiences, the answers to some questions and some new questions.

There are those who dismiss the whole idea that one can study the evolution of behaviour with anything approaching scientific respectability. Bones are one thing but behaviour quite another. You can handle and measure bones. But, as I have tried to show in these pages, one of the chief merits of comparing the behaviour of related animals is that the changes which have taken place since they diverged from ancestral forms become apparent. And change in this sense means evolutionary change. Of course I had no inkling when I started watching gannets that I would see all its relatives as well. But when those opportunities came, it was the gannet which provided the reference points. Take, for instance the gannet display which partners perform when they are changing guard at the nest. I described it earlier and showed that it was not an 'excuse-me' posture, aimed at neighbours, but was directed at the mate to signal the intention of departing from the site. Its most conspicuous features are the strained upward reaching of the head, with the bill pointing skywards, and the curious lift of the wing-tips caused by swivelling the upper-arm at the shoulder joint. Is this skypointing display to be found in the boobies? The answer is simple – yes – but with large and highly significant differences which speak volumes about behaviour theory. It would not be extravagant to say that the following simple observations would *in themselves* establish some important concepts. I hasten to add that the concepts which I have in mind owe nothing to me, but the fact remains that simple behavioural observations of this kind can have implications at that level.

My first skypointing booby was on Tower Island, where a lovely little group of masked boobies, immaculately white birds quite like gannets, nested near our tent. A male was clearly interested in a female who was perched on a lava outcrop beyond his reach. He skypointed repeatedly towards her. Up went his head and wing-tips and out came a thin whistle – 'phee-ooo'. This was skypointing all right but he obviously was not signalling his intention of moving away. On the contrary, he was intent on staying and enticing the female to him, as events showed on this and many subsequent occasions.

Soon afterwards I saw the same display from a diminutive male red-footed booby in a Palo Santo tree. Again, he was signalling to a nearby female but there were some glaring differences. As well as raising the head and wing-tips he tilted the tail skywards, too. And whereas the female masked had simply responded by approaching, the female red-foot displayed back at the male as soon as he had relaxed his posture. And the accompanying call was not a whistle but a tinny, rasping groan. In both species when the females did approach in response to this sexual advertising they were briefly attacked, or sometimes a ritualised sparring bout ensued. This was very like the situation in the gannet when the female responds to the male's soliciting headshake and is then vigorously attacked before the pair begin their meeting ceremony.

Not until we arrived on Hood Island could I see the blue-footed booby's skypointing, but it was worth waiting for. Sure enough it was again a sexual display but this time the posture was truly bizarre. The skypointing part was typical enough but the wing-swivel was so extreme that they looked grotesquely broken, almost wrenched out of their sockets so that the backs of the wings faced forwards in the vertical plane. And whereas in the masked booby it had been male to female, and in the red-footed male to female followed by female to male, the blue-footed boobies displayed simultaneously to each other. The male's whistle was thin and piercing and the female groaned.

So what has been going on during all the millions of years that the gannets and boobies have been separate species? It

White. Sky-pointing, male to female only

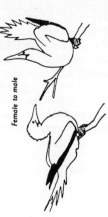

Male to female

Female to male

Brown. Sky-pointing, male to female; also performed in flight

Red-foot. Sky-pointing: unilateral and reciprocal but not mutual

Sexual advertising displays in the Sulidae. In the boobies, the skypointing displays are homologous. The gannet's sexual advertising is merely analogous. Gannets use the homologue of skypointing for another communication function.

Gannet. Head-shake and reach, mainly male to female

Peruvian. Sky-pointing: unilateral reciprocal and mutual

Blue-foot. Sky-pointing: unilateral reciprocal and mutual

seems that in some ancestral sulid, before the gannets and boobies became separate lines, there was a skypointing display. Present-day gannets still have it and so do present-day boobies, but now one of these two lines uses it differently from the ancestor of those early days. This must be so, since gannets use skypointing as a pre-departure signal and boobies as a sexual invitation. And the display itself has changed. But the first point to establish is that the display really is one and the same in the sense that all the present-day versions of it derive from a common source; that the components are basically the same. Fortunately there are at least five components: the neck-stretch, the wing-swivel, the tail-tilt, the foot-raising and the special call and all of these are present in all skypointing boobies, with the partial exception of the foot-raising. That they should all have arisen and been combined fortuitously, anew in each species, is so improbable that it can be dismissed out of hand, the more so because the wing-swivel is a perfectly graded series of changes from the simple tilt of the masked booby to the bizarre version of the blue-foot. There is no earthly reason why this sort of resemblance should be ignored when comparable physical ones are routinely accepted as evidence for relationship.

So it appears that the 'same' display serves two different functions in the family. In the gannet it says 'I am about to leave my site' (or more precisely 'I am feeling like leaving my site', for it is not *always* carried through) and in the boobies it signals 'I am a sexually-motivated male (or female in the species where both sexes use it) inviting you to approach'. Now if the same display serves these two different functions, or to put it another way, sends out these two quite different messages, then it follows that the species which signal 'I am about to leave' *feel different* from those which say 'I am sexually interested'. They have different motivation in the two cases. This is obvious, but to make it even more so consider what it would mean to say that an animal could 'feel like' doing one thing but send out a message, via its display, that it 'felt like' doing something quite other. Apart from special cases of 'deliberate' deception, it would lead to chaos in communication systems.

Skypointing, therefore, has changed not only its form and its message; the original 'feelings' which were tied to the original display have been supplanted or superseded by 'new' feelings associated with the changed form and function. This 'emancipation' is an idea well worth having and as I said earlier, simple field observations can be valuable evidence.

So far, I have merely offered a few verifiable observations and made some justifiable deductions. One could ask many more questions, such as why it was this particular display that changed from 'I am about to leave' in the ancestral sulid to 'I am sexually interested' in the modern booby. But any attempted answer would have to go beyond the evidence. There are many questions that can never be answered and it is well to know which is which. You may have noticed, incidentally, that I never said why it should be assumed that the *original* message of the display, and probably the original form too, was 'I am about to fly', as it still is in the gannet, and the *derived* message and form is the sexual one that is now found in the boobies. Why couldn't it be the other way round? There can be no absolute proof but, as I hinted earlier, the form of the display, and its message in the gannet, is just as one would expect if it were indeed derived from the intention movements of flight which commonly involve neck-lengthening, wing-flicking and foot-movements. By elaboration, these could readily give rise to the new form seen in its most extreme version in the blue-foot.

That was quite a detour, and one that I couldn't have taken but for the gannet. Now, when I see the familiar display on the Bass, I look for things I may have missed. Is it possible that I have misread the whole thing? To me, it seems not, but wouldn't it be interesting if somebody else could come up with a carefully observed and argued alternative case for the whole phenomenon! But I would go as far as to say that if the skypointing story were to be comprehensively overturned on good grounds, then the art of comparative behaviour study would be in some sort of trouble. At another level, more could be gleaned from a really detailed examination of the context, sequence and timing of the movements and of its relationship to the status of the pair-bond.

Other gannet displays did not have such easily recognisable counterparts in the boobies but, because it is obvious that they all need to communicate certain basic 'moods' or 'tendencies', I simply took the main communication-requirements in each species and looked at what actually happened. There was, for instance, the need to communicate the intention to defend territory. I have already said quite a bit about the gannet's version, and how it originated in ground-biting which became 'polished' into a ritualised display. If I expected to find something very like it in the boobies (which would have been naive anyway) I would have been disappointed. But they had to defend their territories, so how did they do it? The red-footed booby gave me the strongest reasons for believing that I was on the right lines with the gannet. The male flew into his tree or bush with a rapid, tinny rattle, the obvious equivalent of the gannet's 'anchor chain running out', which it delivers as it flies in to its site. The red-foot then launches into a series of rapid sideways and downward swings of its head, still calling. Some of the downward swings ended by biting the twigs. It seemed to me that the red-foot was doing much the same thing as the gannet, namely directing its aggression towards something that couldn't bite back – redirected aggression – and just as the gannet's ground biting had become stylised, ending up as smooth bows, so the red-foot's twig biting had ended up as these downward and sideways swinging movements. The gannet's bow was rather more complicated, since it incorporated the headshake, but then the red-foot hadn't been as likely to bring in the headshake because twigs are clean and dry whereas ground is wet and muddy, and the headshake is basically a beak-cleansing movement. As before, the observations are correct and verifiable, and any alternative interpretation that fits them better will be as enthusiastically received by me as by anyone else.

Among the most evocative bird displays are the highly evolved greeting ceremonies which seabirds, more than most, often possess. It may be because in them there is something recognisably akin to our own behaviour after separation. Just as chimpanzees hugging each other are universally recognised, by us, as expressing affection, so a pair of gannets, standing breast to breast, calling loudly and fencing with their bills, are not only demonstrably greeting each other, as the context shows, but are engaged in a display which seems particularly easy for us to understand; it 'clicks'. Whether we are correct is another matter. In fact the gannet's greeting ceremony caused me a few headaches. As usual, the problems became apparent only when I began to 'measure' the display. At first it was simply 'greeting' behaviour. The pair re-united áfter a fairly long separation, 'greeted' each other. But what exactly did I mean by that? When I greet my wife, what do I do and why? (I am not being anthropomorphic here – just asking what the term 'greeting' may hide). If I had just been to the village for an hour I might simply pick up where I'd left off without even a 'hello'. Away for the day would probably lead to a greater expression, but still casual. Returning from a long trip abroad there would be more intense feelings on being re-united. At each level, the behaviour would obviously differ and could easily be measured. In fact I once put a student onto this problem. Discreetly hidden, she logged several aspects of greeting behaviour at various locations, including a railway station (medium separations) and an airport (presumed longer separations). As always in this sort of study, there were many complications which made analysis difficult but, nevertheless, there were measurable differences.

So what did I find in the gannet? First, that long separations were followed by long and 'intense' greeting display. The intensity (apart from the duration) could be measured by the 'strength' and frequency of the various movements. For instance, the wings could be held out a little or a lot; the fencing with the beaks could be rapid and vigorous or more desultory, and so on. Second, that old and experienced partners 'greeted' even more enthusiastically than new pairs; and third, that greetings were at their strongest early and late in the season. In addition there were slight but nonetheless significant differences between male and female in the details of their greeting behaviour. The context of the display also was important, for it was not only a greeting. It occurred very strongly whenever there was a fracas near to a pair, or when a departing bird blundered clumsily past. In fact, any occurrence which would have provoked the territorial (site-ownership) display from a single bird, provoked the greeting display from a pair. Finally, it tended to precede (and often to follow) copulation.

These observations suggested a good deal about the gannet's 'feelings' and about the role or function of the greeting ceremony. Sexual feelings or tendencies seemed obviously involved and could be expected to be stronger after long separations. But aggression seemed just as clearly indicated, especially when the greeting ceremony was performed in response to potential intrusion. In fact the whole display was highly reminiscent, in form, of the territorial display within the constraints provided by two birds 'bowing' simultaneously on the site, and certain other modifications. Even the apparently unrelated observation that the greeting ceremony waned in mid-season supported this interpretation, for displays such as threat or 'bowing', which were obviously aggressive, also waned in mid-season and rose again towards the end.

Perhaps a number of pieces are now falling into place, for earlier I made a big point of the presence and role of aggression in the gannet's pair-relationship. Think of the vigorous nape-biting by the male when the pair meet, and of his sustained gripping and biting during actual copulation. It is not surprising that aggression, as well as sex, should figure prominently in the meeting ceremony, albeit in a modified form. And particularly unsurprising that the pair, when feeling aggressive because of the intrusion of a neighbour, should express it 'on' each other by means of the meeting ceremony. As to the functions of the display, and following the same line that I used in trying to 'explain' nape-biting in copulation, it seems likely to be to form and then to consolidate and strengthen the pair bond. Aggression may be a strong and at first slightly disruptive component, but if it is 'bent' to the service of the pair bond, all the better.

It was with all that I had observed about aggression in the gannet pair context much in mind that I looked at the red-footed boobies behind our tent in Darwin's Bay. The aggression upon meeting couldn't have been clearer. The male red-foot in fact often lost control and pressed home a furious attack. And the female betrayed clear signs of fear, trembling, flinching, ruffing out head feathers and turning her head away from the male. Here, the situation was even more remarkable than in the gannet, for the female red-foot is considerably larger and stronger than her mate and could easily beat him, but she must *not* retaliate if the pair bond is to be formed. Similarly in the masked booby, the greeting ceremony was a very vigorous and hostile-looking 'sparring', in which the opened beaks were flung at each other with an audible clatter. But it was perhaps in Abbott's booby that the greeting ceremony was most interesting of all. As I described earlier, this magnificent seabird nests in the jungle canopy of

247

Christmas Island – a very dangerous habitat. Here, the direct expression of aggression would be hazardous indeed. Abbott's booby overcomes this by conducting its spectacular greeting display at a distance and by strong inhibition of contact-aggression when the pair do meet on the nest for copulation or change-over.

There is still much to be learnt about pair-relations. Of course this is true of all social relationships, defended, as they are, by their complexity and their annoyingly 'fuzzy' dimension which makes them difficult to quantify. Not that I go overboard on quantification – far from it – but I do believe that opinions are only as worthwhile as the evidence that supports them. To return to the Bass. Wouldn't it be interesting if I could tell you for any (preferably every) pair in my observation colony just *how* that pair's greeting behaviour had been conducted on each and every occasion since first they came together, perhaps twenty years ago? If you say 'no', you have no feel for behaviour. What subtle changes it may have shown as the relationship developed. But if you imagine gannets 'conversing' by means of it, you could not be more wrong. There is, as in all its behaviour, that ineluctable element of unthinking adherence to the pre-programmed pattern. Take, as a simple but thought-provoking instance, the female that caught her lower mandible in a piece of netting which has been built into her nest. Her mate arrived out of the throng circling the rock, calling as he approached for landing. As usual, he bit her neck and she faced away. Then, as she had done hundreds if not thousands of times before, she reached up to begin bill-fencing, the meeting ceremony. But the netting dragged her lower mandible down, making her gape in a threatening way. The male immediately attacked her whereupon she appeased him by facing away once again. Then she tried again, but with precisely the same result. And so it went on, time after time. Any fool could see the cause of her behaviour. No doubt her mate could 'see' it; his remarkably acute eyes could scarcely fail to register the netting wrapped around the bill, but not only was he totally unable to 'understand' it, the female could not even communicate the 'simple' fact that something was 'wrong'. They were effectively automatons.

So it would be fascinating to know more about the fine details of communication in gannets, or any other animals –

how it changes with time, experience and context. There are ways in which a start could be made, but they are time-consuming and therefore expensive. I once calculated roughly how many thousands of hours I have spent watching gannets on the Bass. I don't consider any of them to have been wasted. It is said that people come to resemble the animals they identify with. That would make me, like the gannet, handsome, strong, an indomitable fighter, a faithful and long-lived mate – and deeply and ineradicably unintelligent. Homo non-sapiens.

Index

rabbits (on Bass), 26
Rhum, 4, 219

St Andrews University, 7, 8
Saint Kilda, 56, 63, 90, 107
sand-eel, 89
Scar Rocks, 90
sea-lion, territorial and breeding behaviour, 174-7
shag, 39, 108, 114-16
shark, black-tipped, 134, 147, 148
shearwater, Audubon's, 136, 152
shearwater, Manx, 4
sheep (on Bass), 26
Skokkholm, 23
skua, great, 114
social stimulation, 53-5, 170, 224
sociobiology, 163
solar still, 141-2
sprat, 89
Stac an Armin, 90
Stac Lee, 90
'strategies' in birds, 102, 114, 137, 171-2, 180, 181-2

Strathbeg, Loch of, 128
Sula Sgeir, 89
synchronisation and timing of breeding
  in blue-footed booby, 136-7, 180
  in gannet, 54, 86
  in great frigate, 122-3
  see also food and Tropics

tameness in birds, 63, 64
Tantallon Castle, 23
territory and territorial behaviour, see under species
Tower Island (Galapagos), 125, 129-73, 190, 191
tree-nesting in seabirds, 230
tropicbird, red-billed, 172
tropicbird, red-tailed, 208, 217
tropicbird, white (golden bosun), 172-3, 208, 227
Tropics, effects of, on breeding in seabirds, 122-3, 136, 151, 159-61, 180-2

SELECTIVE NAME INDEX